The Theory of Existence
&
The Science of Consciousness

By

Steven S. Sadleir

The Theory of Existence
& the Science of Consciousness
By Steven S. Sadleir

Published & Distributed by Amazon.com via CreateSpace
For

Self Awareness Institute
668 N. Coast Hwy. #417
Laguna Beach, CA 92651
www.SelfAwareness.com
949-355-3249

ISBN: 1478221313

ISBN 13: 9781478221319

Other books by Steven S. Sadleir

Self Realization

Money & Power, the Secret History

Christ Enlightened, the Lost Teaching of Jesus Unveiled

The Awakening, an Evolutional Leap in Human Consciousness

Looking for God, a Seeker's Guide to Religious and Spiritual Groups of the World

Dedicated in loving memory to
Yogiraj Vethathiri Maharishi

Table of Contents

Acknowledgements

Foremost, I want to acknowledge Yogiraj Vethathiri Maharishi for inspiring me to study the nature of existence, the unified force and purpose of existence. His own realization of God and training has moved me to share this awareness with you. I would also like to acknowledge the arch yogi Sri Sri Sri Shivabalayogi Maharaj whose own life force energy enabled me to transcend the confines of mind and matter. I would also like to express my love for God for the inspiration and guidance, and Christ for his loving guidance.

Each of the scientists mentioned in this book should also be acknowledged. I have omitted biographical notations as they would exceed the volume of the text itself. I drew heavily from Wikipedia and other online sources, such as NASA, and acknowledge all those who contribute to these online sources. I thank all my professors and teachers, from each of the disciplines addressed in this book, and particularly Professor Richard Wolfson from Middlebury College and Sean Carroll of California Institute of Technology whose lectures on physics helped me assign names to my experiences, as well as the dozens of other courses taken from the Teaching Company. I also like to acknowledge Rabbi Philip S. Berg for his insights into Biblical theology, and John D "The Judge" Russell for mentoring me in college.

I would also like to acknowledge several of the editors who provided assistance in cleaning up the manuscript, including: Julie Adams, Patricia K. Faust, Richard Fess and Maria Kellis, and to Amazon's CreateSpace team for creating the book.

Preface — How We Perceive

Whenever we look at something, we first need to consider the perspective from which we are viewing it. From where and how we see it, shapes our perception of it, whatever it may be. If your first and only time seeing an elephant is from directly behind it, your perception will be different than if your first and only view is from directly in front of it. The perception a mouse has of an elephant will be different than that of a giraffe or another elephant; and the perception a Masai tribesman in Africa who grows up with elephants will be different from a European who comes to Africa for the first time and sees one. Our perception from only hearing about them, reading about them, or seeing a photograph of one, is quite different from that of actually being with one or interacting with them.

What we see actually often limits our understanding of what something is. Our perception itself limits us. Even our understanding limits our awareness, and our own mind's process of thought can reduce our awareness of what something actually is. Take light as an example, since this is the primary sense most humans depend on for their perception of reality. If you were to graph the scale of light (see figure A below), with gamma rays and x-rays being on the extreme small end of the spectrum and microwaves and radio waves being at the extreme large end of the spectrum, and compare that to the range of visible light a human being can physically see, the visible light is only a few inches on a scale several feet across. Most of what is actually there is not seen.

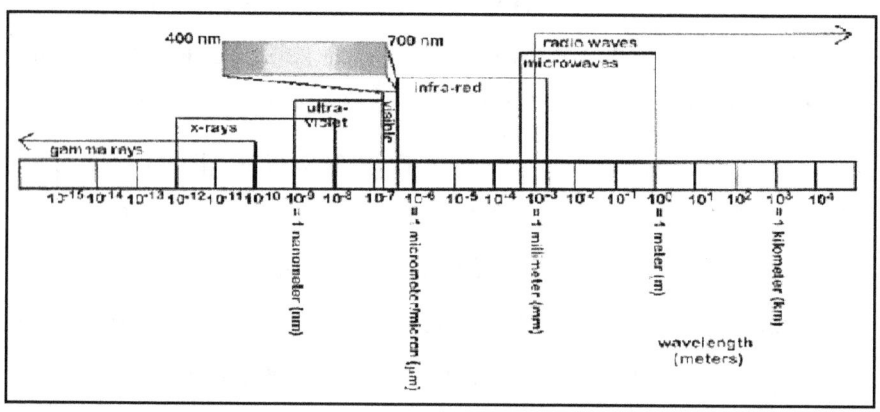

Until the time science could measure the higher and lower frequencies of light, these invisible light rays were not within the realm of human understanding or awareness. We tend to think what we can see with our physical sense perception is what is real, and discount that which appears to be invisible. Therefore, to understand the nature of God, existence or even our own nature, we need to extend our awareness beyond physical sense perception and our own cognitive or mental abilities.

The National Aeronautics and Space Administration (NASA) now looks at space through the spectrums of ultraviolet and infra-red wave lengths, as well as others, and sees other objects and expressions of life different from what we could see through lens telescopes. What we can hear is similarly nothing close to what can be heard through instruments that are calibrated to pick up higher and lower frequencies. Dogs and whales hear sounds that we cannot. Most of what could be seen, heard, felt or even understood lies beyond our physical sense perception.

Moreover, science tends to follows a trajectory of thinking, building upon the ideas of others who have led the way before them. We tend to think about the perspective of another and follow that line of thinking forward, creating a kind of "thought momentum" which leads us in a certain direction of thought. Contrary thoughts and ideas tend to be initially rejected. Once we start thinking along certain lines, we tend to dismiss other possible perspectives or conclusions, which limit our understanding and awareness of whatever we are studying. In order to grasp an idea to its fullest extent, we need to be able to look at an idea from many perspectives including those that might appear to contradict our current understanding.

This redirection of thought is what Dr. Edward De Bono calls lateral thinking. According to Dr. De Bono "lateral thinking is for changing concepts and perceptions." The classic example is "You cannot dig a hole in a different place by digging the same hole deeper." Being able to gain a clearer perception and understanding of anything requires us to reprogram our own brain. Our own perception and thinking can distort the reality of what something is. According to Dr. De Bono:

"The brain as a self-organizing information system forms asymmetric patterns. In such systems there is a mathematical need for moving across patterns. The tools and processes of lateral thinking are designed to achieve such 'lateral' movement. The tools are based on an understanding of self-organizing information systems"

Most people assume that when they see or hear something, others see or hear the same thing, but this is not completely true. Our cognitive understanding taints our perspective and our level of awareness narrows or opens our perception and understanding.

What you already think tends to be superimposed into your reality, and modifies your perception and understanding of an object or occurrence. Each of us experiences life from different perspectives. Thus, we all live in a slightly different reality, even physically.

Newtonian physics operates under a different set of laws than does Quantum physics, and how objects interact at the level of atoms and sub-atomic particles differs from that of planets, stars, galaxies and universes. Moreover, almost all science is only viewing the universe from the level of three dimensions (four with time) and simply discount what lies beyond what we currently know or can physically observe.

Everything that appears arises from what appears to be emptiness. Everything that appears is sustained by forces that are unseen and more often than not unknown. And everything that appears at some point disappears. Everything appearing is only a modification of a previously existing state, and everything that appears is a part of something that is not appearing. All sub-atomic particles appear and disappear. Atoms appear and disappear. Elements, molecules, cells, life forms, planets, stars, galaxies and universes appear to arise out of nothing and appear to go back into nothing. But nothing is the something that everything is appearing out of, is being sustained by, and is dissolving back into.

So as we begin to look at and contemplate the nature of God, the Universe and ourselves, we need to understand the limitations of our own perception, previous notions, beliefs and ideas, and the level of consciousness from which our mind is forming perceptions and ideas. We have to realize that we don't know what it is that we still don't know, and we need to question everything that we think we know. We have to understand from the outset that we really don't know much about God, the Universe or our own nature and purpose. This book serves as a guide to help you move beyond the realm of knowing and understanding and into a realization of existence.

My own awareness of God, the universe and self comes less from learning and more from transcending the mind and senses through deep meditation. I was fortunate to apprentice with the great Kundalini master Vethathiri Maharishi and join in his fields of awareness during meditation, and am a Kundalini master in this lineage. This awareness was deepened during a period of sitting in meditation in India with the arch yogi Shivabalayogi Maharaj for twenty-three hours a day for forty consecutive days and nights, and more recently through deep meditation on existence itself.

If I were to serve you tea in a cup you've been drinking coffee out of, you would have to empty your cup of the coffee to appreciate the taste of the tea. In a similar way, to get a clear idea of what I am going to share with you, you will need to empty your mind of previously held beliefs and perspectives for at least as long as you are reading. Science devotes most of its study to how things are appearing and measuring those changes that physically occur. This study will focus on the underlying causes behind all changes that occur. First we are going to become conscious of what is changing, then we are going to become conscious of the cause of the change, and then we are going to become conscious of the consciousness that is conscious of the change.

Chapter 1 — Introducing Existence

This book is designed to explain the underpinnings of what existence is. An attempt is made to describe: 1) The wonders and workings of that which created existence which we call God; 2) How the physical universe came into being, and what sustains and guides the physical reality as we know it; and 3) Who we innately are, what place we hold in the grand scheme of things, and what purpose we play in existence.

In order to best perceive God, the universe and mankind, I am going to attempt to tie in the underlying science and philosophy that mankind has developed to show the interconnectedness of these unifying principals and perspectives. There are two underlying concepts that will emerge in each chapter: Force, whatever it is that creates change, and Consciousness, represented as the means in which that force takes shape.

God is introduced as that which lies beyond description, yet is a part of everything that is described. The Universe is defined as the physical form in which we see, test and measure in physics and science, with the understanding that there is something beyond the physical properties of existence and our ability to measure it. Man is considered both in the physical sense of a human being with a brain, as well as that which sustains its physical existence and is conscious of its own awareness.

We began our discussion by becoming aware of the very perspective from which we are delving into the discussion and the limitations placed on our own physical senses and cognitive mental processes in the Preface. We need to be aware of our own state of awareness to begin a study of what we are becoming aware of. We must know where we started from in order to gain perspective on where we are going and how to get there. We need to begin with an understanding of who or what is perceiving and what our own perception is.

This first chapter provides the overview of what we are going to be considering. The second chapter attempts to describe what was before the beginning and is after the end. We need to have an inkling of that which underlies existence and the physical universe that we behold. Science focuses on measuring what physically appears, so I will introduce philosophical constructs from the world's great religions, both Eastern and Western, in order to provide a metaphor for that which lies beyond the physical appearance of existence.

The third chapter delves into the very nature of our physical existence. What made the Big Bang boom? What are the underpinnings of physical matter? What brings an energy particle into a physical expression, if only for a fleeting moment? What was the particle before it came into existence, what provides for its existence, and where does it go after its appearance? The particle has no existence without that which provides for its existence, and this is the missing link in modern science. The basis of quantum mechanics and particle physics is discussed in the light of the underlying consciousness that provides for all existence. In other words, we are peering into the mind of God.

The fourth chapter provides a basis for understanding the basic dynamic of physical properties and the formation of the innate intelligence that animates our existence. From the smallest microcosms of

sub-atomic particles and atoms to the greatest macrocosms of galaxies and universes we see fields within
fields of energy interaction between and within each
other. These systems of organizing intelligence provide the great matrix that we behold as life. To understand existence we must understand the workings of
nature, or how the Divine expresses itself through us.

Chapter five begins to layer these philosophical constructs into the working reality and physical properties
of matter and life. Here we are going to explain how
consciousness and energy create life forms -- from the
physical elements that make up the stars and planets,
to the building blocks of life, such as amino acids and
proteins, and how these parts come together to form
a greater whole. There is a guiding light behind the
workings of matter and the formation of organic existence, and to understand this invisible hand is to realize the true nature of ourselves.

The sixth chapter further develops these invisible
properties and characteristics into the wonders of life
that we behold. Here we are going to explore life as
crystalline lattices and biochemistry, and learn the
secret code that directs all life processes. The universe is programmed - and functions according to
prescribed rules and regulations or laws. Life evolved

under these conditions and under this guidance, so to understand our Creator, our universe and ourselves, we must understand these principals and how they work.

Chapter seven brings these natural laws into the relevance of our own existence as human beings, and places the human in the context of nature and life itself. We are going to step back and look at human beings from the greater perspective of life on Earth itself and contextualize the human role in an Earthly existence. Here we discover the invisible hand of God developing and perfecting its own creation, making adaptations and improvements along the way. The inner workings of creation are further explained in the context of a greater interconnected whole.

In chapter eight we focus on the evolution of human existence and the development of civilization. Human beings serve a purpose in existence that they are all too often oblivious to, so in this chapter we are going to contextualize the role of human beings in life as a whole and on Earth in particular. Here we begin to understand the role and purpose we play in existence, and the relevance to our Creator's grand plan. The primary goal of this chapter is to enable the reader to move beyond the self-centered and self-serving conditioning of the mind, to see the greater role and purpose we each play in the grand scheme of things.

The ninth chapter moves us from the world around us into the world within us. Here we explore the development and functions of the human brain. Here is where our perception is generated and our concepts and ideas created and stored. We often think without knowing what our own thought is, how it was formed, and what it consists of. The world we know is created in our mind, yet our own mind omits most of what existence is. Thus, in order to better understand the nature of our Creator, our physical universe and ourselves, we need to develop a greater understanding of how our own brain works.

With a better understanding of what understanding is, we can now go into the philosophical constructs of what existence is. We begin in chapter ten to explore the Eastern perspective of consciousness and the nature of the self. Herein lays a rich treasure of the philosophy on the nature of the self. Here we explore the nature of consciousness, and how to become conscious of our own consciousness. From the ancient yogis, to Buddha, the Jain, and other mystics we transcend the mental concepts and become aware of what awareness itself is and the intrinsic higher nature of our own being.

Chapter eleven brings us back into the realm of science and the Western perspective of consciousness and how it relates to mental functions and sense perception. Here we explore Biblical concepts of man and God and the philosophies of European genius' during the age of enlightenment, and then tie that into the science of psychology today. Here we will integrate the philosophical with the scientific and draw our own conclusions.

The last chapter, twelve, ties all these concepts together to provide a perspective and relevance to our own existence. Here we look at the world as it is today, and from our expanded awareness of the workings of existence determine our role in existence and what courses of action we must now take. We explore both the role and opportunity that awaits an enlightened society, and the destiny that is ours to make manifest. Here we conclude with the purpose of our existence and how to create greater happiness and peace on Earth.

Chapter 2- Absolute Space

To describe anything is to define it, to create a boundary or differentiate it from something else. Neither, God, existence nor the self can be accurately defined or differentiated. There is no boundary to God. There is no beginning or end to that which underlies existence innately, and the self is transcendent to all understanding. What appears to us is superficial, temporal and fleeting, and ultimately unreal. What is real is permanent, undifferentiated and whole within itself. Reality is unchanging, whereas everything in the physical universe and within our senses, conceptions and experiences is constantly changing and being modified, creating a sense of impermanence.

Time, space, energy and matter are relative. From latent potential to patent expression, the play of life is what is projected upon the screen of consciousness itself. Let us use the analogy of a chalkboard to represent the undifferentiated whole, beyond time and space, dimensions or any physical existence. Imagine it as latent potential from which anything can be projected, drawn, or appear. Once you place a dot on the chalkboard, the dot creates an appearance of being separate from the emptiness, like a big bang appearing within the void or static state.

Once you place the dot on the chalkboard, a sense of time and space is projected. In one place there is chalk, but the rest of the board is blank. Therein lies the potential for the dot, but the dot is still just covering the chalkboard. The dot won't stay there long. This is similar to how a cloud appears to form out of nothingness. The elements of oxygen and hydrogen are in the atmosphere but they are not seen. Then as conditions in the atmosphere change, the elements come together, and apparent differentiation occurs as the appearance of a cloud. The clouds can form, they can move around in time and space, they can disappear via rain or snow, but the underlying space that provides for their appearance remains the same.

Let us begin by using classical spiritual philosophy to reflect on the nature of existence and consider some inspired perspectives of how the Universe was created. The Biblical God existed before Creation was created. Something or someone was behind the physical origins of existence, and we call this God. Some personify this Creator; others recognize it or Him as being beyond form and conception. We intuitively sense that something existed before the Big Bang, and something is providing for its continued existence. Since we are a being, we sense that there is a greater being that created us. Some invisible hand seems to be orchestrating the movement of the universe and all the parts within it, including us. In the East we hear of a similar perspective: There was, and is, a latent potential from which all things appear to arise, a consciousness that enables anything to take form, and there is a patent expression taking form and substance. From within emptiness arose change – energy manifested – providing physical form and substance. So I am going to use both the Western and Eastern metaphors to address the dynamics of our physical reality.

The underlying potential, the undifferentiated and unchanging whole from which all things arise is given the name Shiva by the yogis. Shiva represents the consciousness itself, and the energy that appears to arise from within it is referred to as Shakti. This energy can take on many and all forms. However, one key consideration to begin with is realizing that all forms are temporal, fleeting and ultimately disappear again and are thus not ultimately real. Shiva's dance is the creation and destruction of all appearances. Energy particles, atoms, molecules, cells, life forms, planets, galaxies, and universes are coming and going, but what provides for them all remains the same. The underlying consciousness is unfettered, undifferentiated and unchanged. Only in appearance do changes occur, and all changes are only a modification of a previously existing state.

Coming back to the chalkboard analogy, any projection upon the chalkboard that you make will vary to some degree. The projection itself puts pressure on the chalkboard, causing the chalk to stay longer or shorter in duration or in quantity or quality of appearance. This very projection upon the chalkboard creates a sense of time and space. At one point the chalk may be applied on the upper left and then move to the right, relative to the perception of consciousness from which it is viewed. The chalkboard itself remains the same regardless of what is projected upon it, but our attention is drawn to whatever is being drawn or appearing.

As we draw lines upon the chalkboard our awareness is constricted to a linear perspective and time and space appear to occur. As we move our hand from left to right, the sense of there being a then and now appears, and the projection of the hand moving across the board creates the sense of what will or may be in the future. The reflection of that potential creates the manifestation of that appearance; this self-reflective quality of consciousness is what makes the manifestation of this latent potential possible.

Sooner or later the chalk will fall away but the chalkboard itself has not changed; only what has been projected upon it changes. The universe and everything within it, is projected; and it is moving, expanding outward, but the consciousness that provides for its appearance isn't changing. From latent potential to patent expression, what comes to arise is a reflection of the consciousness creating apparent differentiation in time and space.

In the Kabbalah, the esoteric teachings within the Jewish Biblical tradition, the universe is described in a similar way. According to the *Sefer Yetzirah*, or *Book of Creation*, which is the oldest of the Kabbalistic texts: "From the infinite, undifferentiated, potentiality that is God's light, the universe was created by removing the light, like 'hollowing out a pumpkin'." According to the

Tree of Life, God's light was too bright for us to see in its totality, and so only one small thread of light as fine as a hair came into our perception as the living universe. We are told that our own true nature is but a "spark of divinity" that has been rooted in ignorance of our true nature, and that life is the process of allowing more light to shine through us through self-reflection and direct awareness.

The Creator, Yah or Yahweh's, light comes into the universe like a rainbow of colors from one white light into what are referred to as the ten emanations or *Sefirot*, and these points of light guide us to attain enlightenment or God realization. This Divine Consciousness is rooted or grounded in the physically manifest world, what is called Malchut in Hebrew. From this most base awareness we evolve, or follow the ascendance of light back to the Source itself – Yahweh or God. Our life experiences cause us to develop greater awareness as we transcend, expanding our own consciousness until we reach full consciousness, represented by the Hebrew word Keter or Crown, which refers to the enlightenment of the consciousness. God realization is where the individual consciousness is undifferentiated, and complete or total consciousness is realized. The light of our own indwelling spirit displaces the darkness or ignorance of God's living presence. Thus, from our fall from grace, we come back into the light and live in peace through God awareness or enlightenment.

In another Kabbalistic text called the *Bahir* the creation of the universe is described as follows:

"Before all things were created...the Supernal Light was simple, and it filled all Existence. There was no empty space...When His simple Will decided to create all universes...He constricted the Light to the sides...leaving a vacant space...This space was perfectly round. After this constriction took place...there was a place in which all things could be created...He then drew a single straight thread from the Infinite Light...and brought it into that vacant space...it was through that line that the Infinite Light was brought down below."

According to Rabbi Philip S. Berg's translation of the Kabbalistic text called the *Zohar*:

"*Zohar*, our Holy Grail, leads us to a state of mind in which we are connected with the Infinite continuum, where time, space, motion are unified, where past, present and future are entwined, where everyone and everything is interconnected, where then is now, and now is beautiful."

The *Zohar* is used to decipher the meaning of the metaphors used in the *Bible*, and in Beresheet (A, 2) it is written:

"When God created the world, He knew that we, the vessels, could not receive His awesome blazing Light in its totality. This light is revealed in measure with direct proportion to the degree of change we've undergone, allowing us to receive a greater portion of hidden light."

God's light is represented as consciousness, a living presence that is omnipresent, and to the degree that we raise our consciousness we become present to God's presence and realize God, the nature of existence and the nature of our self. Our ignorance of our true nature constricts our awareness of God and the universe, but as we become more conscious our individual consciousness merges into total consciousness, or more accurately, the illusion of being separate from God dissipates like the fog of unknowing in the light of the Sun.

As we enlighten, the space is filled with light and we realize the true nature of God, of our existence and our true nature. Only then will our purpose in existence be realized. This principal is also revealed by Jesus Christ in the *New Testament* when he states: "Me and my Father are one." That is, they are one in consciousness, of the same substance or of the same light. He had enlightened. Thus, he saw the Father in everything and realized that "the Kingdom of Heaven lies within." In other words, the presence of God or consciousness of the Divine was within us, and within everything, everywhere, all the time.

We tend to see the universe as a vacuum or void, because our own awareness is constricted to a linear perspective caused by our dependence on our mind and senses. Within and beyond the physical universe the absolute space is filled endlessly with consciousness itself, the latent potential from which all things appear to arise from, are sustained by, and ultimately merge back into. This latent potential has been described as plenum, ether, static state, zero point or unified force, but even those are only expressions of this innate potential. This primordial consciousness in expression is life as we know it, similar to how the consciousness within us is expressed, physically, through what we think, and say, and do.

Inherent in this primordial and infinite conscious-ness lays the latent potential and patent expression, consciousness and energy, Shiva and Shakti. These attributes are the underlying aspects of all creation. All differentiation is only an appearance, all things that appear will disappear, but the Consciousness itself remains unchanged, undifferentiated and whole within itself. God, the Creator and sustainer of the universe, is as it is, and only our perceptions and understanding of God changes. The universe changes only in appearance, not in essence. We are of the same essence and only change in appearance. This is what is realized when our individual consciousness is unbounded by the projection of appearances.

Consciousness existed prior to existence. It has no beginning or end. It never was nor will be, it just is. This primordial omnipresent presence is what provides for all that appears to exist, and long after existence exists the consciousness will remain as it is undifferentiated, complete, and whole within itself unchanged or changing. Appearances come and go, but the consciousness remains as it is. Our perception or understanding of it, or of any of its expressions may appear to change, but it remains as it is. To understand the nature of the physical universe, we first need to understand what underlies it. We must become conscious, of what consciousness is.

Within and around the universe the absolute space is filled endlessly. We imagine it to be a vacuum, but it is not. This apparent nothingness is the "something-ness" that underlies everything that appears to be. The opening verse of the *Bible* states: "In the beginning God created the heavens and the Earth". So even before the heavens were created there was God. Thus God is the underlying consciousness that was conscious to create creation to begin with. Heaven and Earth were created, from a latent potential to patent expression. Thus something infinitely more powerful than creation itself, provided for creation to be created.

The *Bible* goes on to say that "the earth was without form, and void; and darkness was on the face of the deep. And the Spirit of God was hovering over the face of the waters." Whether you are religious or not, this account helps us understand the principal of creation and existence. Something arose that was without form, void of matter or mass; no energy or light had come into appearance. Then the spirit of God, an expression of consciousness, was hovering. The expression of consciousness was the precursor to the manifestation of being. Something invisible brought about creation. Something invisible animated our existence. We call this spirit. Consciousness in expression is spirit, God expresses itself through spirit.

In the East we hear a similar account of how Lord Shiva created existence through his Shakti or divine energy. Consciousness in expression, or what we call spirit, brought about our physical existence. So we have two fundamental components to our existence given from both the Eastern and Western traditions: consciousness and energy. There are two inseparable and inherent characteristics of the absolute space: force and consciousness. Force is the fluctuating, moving and changing phenomenon and consciousness the processional functioning order arising from the reflection of its expression. Everything appearing is an expression of God's consciousness; the expression of consciousness as life force is life itself.

Chapter 3 – Life Force Energy

Albert Einstein once said "Nothing happens until something moves." Everything we experience is only a modification of a previously existing state. Energy is change; it is consciousness in expression in one state or another, changing from one state to another. All life as we experience it involves our mind and senses observing states of energy of one type or another moving, changing, expressing in the multiplicity which is the physical universe as we know it.

From the smallest microcosm of energy within the interactions of sub-atomic particles forming atoms, molecules and mass in elements and life forms, to the greatest expressions of life force such as stars, galaxies and universes, we are observing changes in energy or life force. Like looking at the surface of the ocean, we are observing the ripples, currents and tides of consciousness, but we are only seeing what is occurring upon the surface of our existence. We are only observing the change, not the changeless.

The laws of nature are measuring changes in life force we call movement, gravity, magnetism, radiation, heat and light and many other forms. All physical things are made of mass, comprising atoms, which are made of energy particles or waves. But energy particles are merely what is appearing at any given moment of time and space, and are, in fact, connected and a part of what precedes their appearance and continues on after their appearance. Scientists refer to these links as waves or strings, but all the waves and strings are only modifications of the consciousness which under-lies them and they are not separate from the medium through which they appear.

Our physical senses are taking in different expres-sions of this life force as light through our eyes, sound through our ears, heat and pressure through our skin, chemical interactions through our sense of taste and smell. All that we think is simply energy moving through fields of neurons in our brain, our emotions comprise energy moving through our brain and the reaction of chemicals those energy impulses create. When we think, we are expressing energy. When we speak or act, this is an expression of consciousness that shows up as energy moving through and out of us. What we innately are, what we think, say and do, are all expressions of this consciousness as energy.

The ancient yogis of India realized that everything was innately consciousness, and as this consciousness was expressed we experienced it as energy – Shakti. They believed the entire human body was made up of energy particles so small we could not see them called "Anu" derived from the first name of God, in both the Sumerian and Harappan cultures - "An". Later this primordial Deity from which everything arose was called Anu, again in both Akkadian (who conquered the Sumerians) as well as in Hindu culture. In the other Semitic language of Arabic they enunciate An as Al as in Allah, "the God." In Canaanite Al was enunciated as El as in El Shaddai (God of the Mountain), the term the *Bible* first uses to refer to the one God. In Egypt this primordial Creator was called Ani, as in Horus Ani.

Later the Semitic (Kenite) tribes referred to this one God of all creation as Yah or Yahweh, and in the Hindu civilization the name of the Creator transliterated from An to Anu and eventually to Siva or Shiva. But they all recognized the same one infinite and inconceivable Creator of Creation, and they all recognized creation as being an expression of God. In the East they refer to this life force expression as Shakti, in the West it's called spirit or Holy Spirit. In science it's simply called energy.

In Western culture we first hear of energy described by the Greeks as *energeia*. The philosopher Aristotle uses the term in the 4th century BCE. The concept was later developed by German philosopher Gottfried Leibniz (1646-1716) as *vis viva* or living force, and he recognized motion, or energy, as a constituent part of matter. English scientist Sir Isaac Newton (1642-1727) further developed this concept and developed laws to define and measure it, such as calculus. In 1807 English scientist Thomas Young (1775-1829) was the first on record to start using the term energy in its general use today.

Generally, energy is defined as "a force acting through a distance". It is seen as a force that exerts a pull or push against a basic force in nature along a path of a certain length. Energy is whatever appears to change. Physicists recognize mass as energy too. When matter (energy particles) are changed into energy, such as motion or radiation, the mass of the system does not change -- only its characteristics or expressions do.

According to the law of conservation of energy, which works within the dimensions of time, space, energy and matter, energy can neither be created nor destroyed by itself. It can only be transformed. Everything that exists physically is just energy changing from one form to another. This is Shiva's dance.

We observe this energy transformation through all natural phenomena. In the context of chemistry, energy is an attribute of sub atomic and atomic, molecular or aggregate structures. Chemical transformations and reactions are the result of a change in one or more of these structures, accompanied by an increase or decrease of energy of the substances involved. Protons, neutrons and electrons, as well as acid and alkaline, are examples that these positive and negative forces take different characters as energy.

In biology, energy is an attribute of all biological systems from the smallest living organisms to the biosphere itself. Within an organism, the life force is responsible for growth and development of the biological cell or the organelle of a biological organism. Energy is often said to be stored in cells, in the structure of molecules of substances such as carbohydrates (like sugars), lipids (fats) and proteins. This life force energy is what animates the cells existence.

In geology we observe life force energy being expressed as continental drift, the creation of mountain ranges and volcanos, and in the manifestation of earthquakes, tides, and the very electromagnetism being generated from the Earth's core and radiating around the planet. Meteorological phenomena like wind, rain, hail, snow, lightening, tornadoes and hurricanes are the result of this energy transformation brought about by the solar energy interacting with the atmosphere of the Earth.

Cosmologically and astronomically the phenomena of stars, nova, supernova, quasars and gamma ray bursts are the physical universe's highest outbursts of energy transformations of matter. All stellar phenomena are driven by various kinds of energy transformations. Both gravitational collapse of matter and nuclear fusion provide the basic dynamic of stellar cosmology. The Big Bang itself is merely consciousness being expressed, creating the dynamic of outer expansion and inner contraction creating the three dimensional (four with time) universe that we observe.

Any focalization of consciousness creates apparent differentiation. Like the dot placed on the chalkboard, its appearance is relative to all the other places where it does not appear. Its appearance is only a modification of a previously existing state, and its appearance is provided by, sustained by and merges back into other aspects of its own nature as consciousness. From latent potential to patent expression, consciousness reflecting upon itself creates form and substance.

Like whirlpools created by the tide, consciousness moving in upon itself create an appearance of something distinct from within itself, but it is ultimately unchanged in its essence or substance. The whirlpool is still just water in motion; it is, in essence, the same as the water supporting its existence. The change, or motion, creates an apparent differentiation. The medium for energy is consciousness, and it has within it the self-reflective or self-rotative force that creates apparent differentiation or form.

Like a thought that has a potential to be expressed, iIts potential was latent within the stored memory of the mind itself; it is held within the consciousness. The thought appears when reflected upon --upon self-reflection. The thought was innate within the consciousness and only appeared as something new when the consciousness was delineated. Consciousness moving linearly through the mind creates thought. There was the potential for the thought before it arose, but it was only realized when reflected upon. When the consciousness was reflected back to the origin the thought arose. The change appeared when the awareness was reflected back to the origin of the one thinking. The thinking was only an expression of consciousness, its appearance relative to the one creating the thought. This self-reflective nature of consciousness creates a focalization of consciousness. This self-rotative force creates a sense of I in the consciousness that appears to differentiate itself from the consciousness itself.

This self-reflective nature of consciousness works in humans as awareness creating our reality, just as the self-reflective nature of consciousness itself creates Creation. Any point of consciousness arising has inherent within it a self-rotative, or self-reflective force. The self-rotative force within matter is thought of as a particle or wave function. The particle is only part of the cycle, the part expressing that contrast to the parts that are not expressing or creating apparent differentiation. Each energy particle has a self-rotative force; they spin. Each sub atomic particle has a self-rotative force; they spin. Every atom has a self-rotative force. The Earth, the Sun, and the Galaxy have a self-rotative force; they spin. As each does, they are creating. The universe itself has a self-rotative force. Everything in physical existence is moving.

When a wheel rotates, the speed at the periphery will be greater than the speed we observed in the middle. There is a point in the middle where there is no motion. This point is the static state. This static state is covered or hidden by the dynamic motion around it. As the universe expands the continuous pressure of the static state on all sides of the universe penetrates and affects all parts within, including the static state within all particles, causing them to expand and spin. When you heat water, it forms bubbles and they move. In simple terms energy causes an expansion and movement from within.

Visualize the universe like an extremely large ball or balloon, full of energy particles, and their associations as various types of masses. All the sides of this huge ball are being surrounded by the static state, and the whole universe is being pressed. As the static state is all penetrative, all the energy particles and the static state within them are affected. The static state within each energy particle is connected with the static state around the universe. Nothing is aloof or separate from one another. Thus, one is in everything and all things in one.

Whatever the size or volume of a rotative motion may be, there will be a point in the middle where there is no motion. The static state exists in the middle of every particle. The static force all around the particle by its own mighty pressure, penetrates the middle point in all directions, and this becomes the gravitational force in the particle. The middle point of the particle is the gravitational force which is what attracts the energy of other mass creating atoms, elements, planets and stars. The continuous pressure of the static state causes the middle point of the particle to expand in volume creating a wave or whirlpool effect, which is the energy particle.

According to the intensity of the wave and the distance between one another, and the receptive and reflective quality of the particles or mass, the character will spread from one to another. Just as waves reshape the surface of water, this wave creates the appearance of a curvature in space-time that Albert Einstein used to describe gravity. Wherever the energy is not pushing, the relative lack of energy comes in towards the source like a pulling. Thus objects not subjected to pushing forces move in the straightest possible paths in this wave curvature in space-time. Everything created will ultimately come back to its source, these are called tidal forces.

The gravitational force is conditioned by the all-round pressure of the static state (consciousness)

in the particle, in the mass, and in the whole universe. Expansive forces from within are pushing out,
as contractive forces are pushing in from around, like
the tides of creation. The knowledge of this interplay
between gravitational force and the repulsive force
caused by the spreading wave of energy particles help
tie the loose ends in the sciences of physics, biology
and psychology.

As the consciousness reflects back upon itself apparent
differentiation appears to occur, like a whirlpool. For
everything that is occurring there is a place it appears
to come from and to go to, but the two are part of the
same system. This change appears as: Expansion and
contraction, centripetal and centrifugal, positive and
negative, yang and yin, Shakti and Shiva. The energy
particles go round and round, creating a system, a cycle,
or rhythm. All expressions of consciousness appear
with pattern, precision and regularity. The expression
of energy is structured. This expression of consciousness forms the basis of intelligence in the universe.

The movement from one state to another state forms
a relationship of connectivity we think of as a field.
The two limits of the system work together to form
a whole. That interconnectedness is the field of consciousness that defines the system. Within an atom

we see this system represented as a proton-neutron combination and an electron. There is a nucleus in the center of the atom and the electron field around it. In the Earth there is a core and a field of electro-magnetism around the Earth. There is a field around each atom, around each molecule, around each cell, each life form, each planet, each star, each galaxy, and each universe.

There are fields within fields within fields, from the smallest microcosm of energy particles and atoms, to the largest macrocosms of suns, galaxies and uni-verses - each interconnected with the others, each dependent upon the other, each affecting each other as a part of a greater whole. Consciousness appears to be stratified depending upon the focalization of consciousness of the perceiver, each is a part of itself. Our own consciousness is a part of the cells, mole-cules and atoms within our body as much as with the Earth and solar system, galaxy and universe around us. All consciousness is interconnected and part of a greater interconnected whole of being, but the con-sciousness itself never changes only the expressions of it appearing become defined upon self-reflection.

Every energy particle is spinning. Every atom is spin-ning. The energy around a molecule is spinning. The field around a cell is spinning. The field around a life form, including a human, is spinning. The Earth is spinning around and around, it is also wobbling around and around, as well as spinning around the Sun. The Sun and the earth and other planets are all spinning around the galaxy. The galaxies are all spinning. The universe itself is spinning. Everything is expanding and contracting; everything is spinning and creating waves. The energy particles are pushing out and the static state is pushing in. The force pushing in is creating the waves pushing out. From yang to yin, and yin to yang, from Shiva to Shakti and back again, everything is self-reflecting, moving, and changing within itself.

We think of the atom as the basic unit of matter. It consists of a dense central nucleus surrounded by a cloud of negatively charged electrons. The atomic nucleus contains a mix of positively charged protons and electrically neutral neutrons (except for hydrogen-1 which is stable with no neutrons). The electrons of an atom are bound to the nucleus by the electromagnetic force. This force is the field from within which the pattern operates; defining the system we call an atom. Another way to view it is probability amplitude of occurrence given the system of energy set forth by

the organizing intelligence or expression of consciousness. This oscillation is like a rhythm of energy, the
sound of the microcosmic world.

The electrons of an atom are bound to the nucleus
by the electromagnetic force. Similarly a group of
atoms can remain bound to each other to form a
greater system of organizing intelligence which we
call a molecule or ion. There are relative degrees of
positive and negative energy within each system.
Attractive and repulsive forces working together are
forming systems, with pattern, precision and regularity. Forces pushing inward and outward, from
within and from outside, are forming whirlpools of
energy within the consciousness.

An atom containing an equal number of protons
(positive charge) and electrons (negative charge) is
electrically neutral. The atom has a positive charge
if there are fewer electrons, or negatively charged if
there are more electrons. A positively or negatively
charged atom is known as an ion. Atoms are classified by the number of protons and neutrons that are
in its nucleus; the number of protons determines the
chemical element, and the number of neutrons determines the isotope of the element (its charge). The
vast majority of the mass of the atom is in the protons

and neutrons, over 99.94%. These relative attractive and repulsive, positive and negative, charges determine the character of the atom.

The electrons determine the chemical properties of the element, and influence the atom's magnetic properties. Electrons that are bound to atoms possess a set of stable energy levels, or orbitals, and can undergo transitions between them by absorbing or emitting photons. Photons are an expression of the energy arising from a change in the energy manifesting. Thus, the electron field gives the atom a character that determines how the system of organizing intelligence interacts with other atoms, elements or the energy in its environment.

The idea of systems of organizing energy existing within matter originally comes from the Ajivika, Jain and Carvaka yoga schools in India which date back to the 6th century BCE. The Greek philosopher Democritus coined the term *atomos* in approximately 450 BCE. In 1661 Robert Boyle argued that matter was composed of various combinations of "corpuscles" or atoms, and this theory was further developed by Isaac Newton in the 1670's to develop his theory of light. English naturalist John Dalton proposed that atoms can join together to create chemical compounds in 1805, and in

1827 Robert Brown discovered a motion within water molecules caused by thermal motion that was termed "Brownian motion." In 1905 Albert Einstein produced the first mathematical analysis of this motion.

In 1897 physicist J.J. Thomson discovered the electron, and concluded that they were a part of every atom. Radiochemist Frederick Soddy discovered that there appeared to be more than one type of atom at each position on the periodic table in 1913. Also in 1913 physicist Niels Bohr suggested that electrons were confined into clearly defined, quantized orbits, and could jump between these, but could not freely spiral inward or outward from their system. In 1938 chemist Otto Hahn directed neutrons onto uranium atoms and created the element barium as a product, and a year later Lise Meitner and Otto Frisch, created the first nuclear fission.

In the 1950's scientists began working with particle accelerators and detectors and discovered that atoms were in fact made of systems of energy called hadrons, composed of even smaller sub-particles called quarks. This led to what is called the Standard Model of physics, where both protons and neutrons are composed of elementary particles called quarks. These are further categorized as up quarks and down quarks which are held together by a strong nuclear force mediated by gluons, which are a type of boson which essentially mediates these physical forces between the subatomic particles.

Particles arise within the subatomic field as physicist Peter Higgs has suggested, but those particles are still only a dimension of an interconnected field, and if you dissected the boson particle further there is yet another tier of potentiality arising from within it, and these divisions go on infinitely. The point being, that atoms or any particles within them are comprised of even smaller systems of organizing energy, with attractive and repulsive forces creating a field, and all fields are interconnected as one consciousness. Everything manifesting is energy, which is an expression of consciousness.

All these elementary particles possess an intrinsic quantum mechanical property known as spin. This is analogous to the angular momentum of an object that is spinning around its center of mass. The electrons, protons and neutrons all have a spinning force. In an atom, electrons in motion around the nucleus possess orbital angular momentum in addition to their spin, while the nucleus itself possesses angular momentum due to its nuclear spin.

The magnetic field produced by an atom, what is called its magnetic moment, is determined by these various forms of angular momentum, just as a rotating charged object produces a magnetic field. Similarly celestial spheres have orbiting bodies around them. Sub-atomic particles spin. Atoms spin. All mass spins. This spin creates a field, and these fields express the consciousness into greater systems of organizing intelligence.

When an electron is bound to an atom, it has a potential energy that is inversely proportional to its distance from the nucleus. Similar to how an object orbiting around the Earth has a greater attractive field the closer to Earth it is. The outermost electron shell of an atom in its uncombined state is known as the valence shell, and the electrons in that shell are called valence electrons. The number of valence electrons determines the bonding behavior with other atoms. In other words there is a relatively greater or lesser attractive or repulsive charge which determines how that atom will interact with other atoms or elements in its environment.

This development of relative and attractive forces within the field of organizing consciousness forms the rudiments of innate intelligence. The field holds the patterns. The patterns give the system its character. Its character determines how the atom will associate with other atoms, elements and forces within its environments. Consciousness seeks expression through ever greater, and increasingly more complex, systems of organizing intelligence. All of life is an expression of consciousness, and consciousness evolves to higher configurations of self-expression.

Chapter 4 – The Laws of Magnetism

Any point in time and space arises from a place that did not appear to begin with. The energy that causes the apparent separation creates a current, a charge. There is an expansion or separation from the origin. Like a wave rippling across the ocean or in a pond, a spreading wave appears upon the surface. The ocean remains the ocean, complete and whole within itself, but the movement of the current creates a ripple that creates apparent differentiation upon the surface.

The whirling motion of the life force, the energy particle or particles, creates a spreading wave. This spreading wave is the dynamic force, the expansive force. This wave penetrates the Absolute space, creating ripples upon the surface of consciousness which forms the three dimensional form of existence, and over time the fourth dimension. From the surface you don't see what lies under the wave, the ocean of consciousness itself is infinite and trans-dimensional.

The joint manifestation of dynamic force against the static force evolves as magnetism. One force moving outward to positive, and the attractive force from which it came drawing it inward to negative. The static force is pushing in from all around the universe penetrating into every particle arising within it, which in turn creates a spreading wave, whirlpool or particle coming out of it. The field created between the two has within it the pattern creating the wave form. The field is part positive and part negative. Like in a magnet, there is a positive force on one end of the magnet, and a negative force on the other, and a field of energy exists between the two, and a current of energy flows within and around each system. This is the basic function of all matter and energy in the three dimensional universe.

In magnetism, the static force is attractive and the dynamic force is repulsive. The result of the spreading wave in nature creates five basic functions or dynamics: clash, reflection, refraction, penetration and interaction. These become the fundamental actions of all matter in the universe which create various forms and conditions in life. In other words, energy interacting with other forms of energy creates the dynamic we call life. Without magnetism there is no movement, there is no change, there is no manifestation...there is no life.

According to the intensity and character of atoms in any mass, the repulsive and attractive forces give effect as multifarious reflections and results. Within living organisms this becomes bio-magnetism. Magnetism lies within each energy particle, each atom, each molecule and cell, each life form, each planet, sun, galaxy and universe. This attraction and repulsion and a field within them is manifesting as energy and matter of one sort or another.

Matter is just another form of energy and consciousness working together. Matter is made of energy particles and directed by a field of organizing intelligence. Intelligence is the articulation of consciousness in expression. Another way to put it is the expression of consciousness through increasingly more developed patterns, held within the field created, develops what we call intelligence. Intelligence evolves as the consciousness is able to develop more articulated forms of expression. The fields hold the patterns, and the attractive and repulsive forces interacting and self reflecting (like a circuit) provides the means for increasingly more articulated patterns to evolve. Intelligence evolves as the consciousness is expressed through a myriad of conditions and experiences.

Science defines magnetism as a property of materials that respond to an applied magnetism field. All materials are influenced by varying degrees by the presence of a magnetic field. Some are attracted to a magnetic field (paramagnetism); others are repulsed by a magnetic field (diamagnetism); others have a much more complex relationship with an applied magnetic field, such as the spin glass behavior and antiferromagnetism. The magnetic state or phase of the material depends on the temperature, pressure, the applied magnetic field or other forces.

Fields of energy have been discussed by the yogis of ancient India, and the Egyptians, but the Western discussion is first recorded by the Greek philosopher Thales of Miletus who lived from 625 to 545 BCE, who, in turn, influenced Aristotle. The ancient Indians and Chinese used magnetism and electric current for healing. The first person to write about using a magnetic compass for navigating was the Chinese scientist Shen Kuo who lived 1031-1095 CE. Alexander Neckham was the first European to describe a compass in 1187.

In 1600 William Gilbert wrote that the Earth itself was magnetic, and deduced that this is why a compass needle points north. In 1819 Hans Christian Oersted in Copenhagen discovered the relationship between electricity and magnetic fields. In 1831 Michael Faraday discovered how magnetic flux could induce a voltage; and later James Clerk Maxwell expanded these insights into what is referred to as Maxwell's equations; and in 1905 Albert Einstein used these laws in motivating his theory of special relativity. In the 21st century these theories have evolved into gauge theory, quantum electrodynamics, electroweak theory, the standard model, as well as the zero point energy, string theory, M theory, the unified force theory, and Steven Hawkin's debate on the Theory of Everything, as well as others.

The basic principal is that moving electric charges create magnetic fields and that particles have inherent spin. In magnetic materials the sources of magnetism are the electrons' orbital angular motion around the nucleus, and the electron's intrinsic magnetic movement (electron magnetic dipole moment).The other sources of magnetism are the nuclear magnetic moments of the nuclei in the material. These nuclei forces are thousands of times smaller than the electrons' magnetic moments, but they play an important role in the nuclear magnetic resonance.

Normally, the electrons in a material are arranged in such a way that their magnetic moments cancel out, this is often due to the tendency of electrons to combine in pairs with opposite intrinsic magnetic moments (+ and −). The magnetic behavior of a material depends on its structure; particularly its electron configuration and temperature. At higher temperatures random thermal motion makes it difficult for electrons to maintain their alignment. The alignment of the electrons creates the fields of magnetism.

Energy particles are either attracted to or repulsed by the other attractive and repulsive forces that they are exposed to. Like working with magnets, put one end of a magnet to another and the two push away from each other. Flip one magnet around and the two come together. As magnetic fields combine they form a "domain" and share the same field. The paperclips in a magnetic paperclip holder will stick to the magnet in the paperclip holder because the fields of magnetism in the metal paperclips align with the field in the magnet. They combine in a common field and will begin to act as one. Each is a part of the same field, a collective consciousness.

This fundamental principal of energies, and matter, aligning to their fields in which they are operating forms the building blocks of all life. This self organizing energy field creates the systems of organizing intelligence that create structures of mass and life forms. Systems are formed by the attraction or repulsion fields aligning themselves in various configurations. Different combinations of energy interactions create various characteristics in substances and environments.

Sub-atomic particles align themselves by attractive and repulsive forces. There are positive and negative charges of energy operating within a field. The energy reflects within the consciousness. In other words, the energy is self-reflective. This self-reflective quality of the consciousness creates the basic building blocks of life. Like spinning your finger in a pool or tub, the motion of your finger spinning around creates a whirlpool. The whirlpool then appears to be something different than the medium of the water that the whirlpool resides in. The whirlpool is still just the water in motion. Similarly all form and substance is consciousness interacting upon itself creating whirlpools of energy we call particles. Particles all have spin, all mass has spin.

Atoms comprise both an expansive or dynamic force that pushes out and a contractive or static force drawing the energy back in. In an atom we see two types of energy interacting: expansive and contractive, centripetal and centrifugal, yang and yin; the electrons and the protons acting together as one unit. The interaction of the energy self-reflecting, moving outward and back upon itself, creates a field. The field holds the pattern that defines the character of the atom. Each type of atom has its own character. In other words it acts with a particular and unique pattern or manner.

A current of energy runs through each atom. Each atom has a negative and positive energy, and each has a field, and a way it acts and interacts based on the patterns held within its field. The consciousness within the atom expresses in a precise way. It holds within its self-reflective nature a system of organizing intelligence. Consciousness is expressed within the patterns being formed within the atom, which, in turn, expresses the consciousness in an articulated way and forms the basis of intelligence. So there is a very rudimentary form of intelligence being expressed within the atom.

Each atomic structure is a unique expression of the consciousness. Atoms interacting with each other interface through attractive and repulsive forces. Some atoms attract to other atoms or their parts, such as electrons bonding and forming new systems, and some react through repulsive forces causing reactions. When atoms bond together and work as a system, science calls these systems elements, or chemical elements. Atomic structures of two or more atoms held together by covalent bonds, that are electrically neutral, are called "molecules", and those with an electrical charge are called "ions". When two distinct elements are chemically combined, with the atoms held together by chemical bonds, those systems are called compounds.

The most abundant elements in the universe are hydrogen and helium. These simplest forms of atomic elements are the result of the primordial expansive and contractive forces science calls the Big Bang. As the gravity of hydrogen and helium builds and gathers together it forms a star. The pressure and heat building within the mass creates nuclear fusion. This is what causes the light, heat and radiation we associate with stars.

As these energy particles get smashed together within the stars they also get fused together which creates the heavier elements of the universe. They are heavier because the atoms within them contain more energy particles. When these massive stars explode creating what is called Super Nova's, they disperse these elements throughout neighboring star nebula. The nebulae of neighboring stars are then filled with both lighter and heavier elements which coalesce into either the solid planets and moons that contain more of the heavier elements or the gaseous planets that are comprised of more of the lighter elements.

To date science has identified 118 elements. Of these 98 are known to occur naturally on Earth. 80 are stable, while the others are radioactive, decaying into lighter elements over various timescales. Those elements that do not occur naturally on Earth have been synthetically produced by man-made nuclear reactions. Hydrogen and helium are by far the most abundant elements on Earth, and oxygen is the most common element in the Earth's crust and atmosphere.

Although all known chemical matter is made up of these basic elements, matter itself constitutes only around 15% of the universe itself. Thus only a relatively small amount of the potential of consciousness is manifesting in a form or a substance. The remainder is a latent potential with forces conducting and supporting all that is being expressed, what science calls "dark matter." Invisible to the eye, it provides the space for everything that manifests to manifest, like the chalkboard providing the space for the chalk to appear.

When one atom connects with another, they can form a symbiotic relationship that further evolves the consciousness expressing through them. These systems of organizing intelligence forming elements and compounds comprise the matter and mass of the physical universe. Thus, everything is consciousness in expression. These systems of self organizing intelligence working together with other systems create increasingly more complex or articulated systems.

When the element of hydrogen combines with the element of oxygen the molecule H_2O is created. Hydrogen acting alone has one character; oxygen acting alone has its own character, but when they combine and work together they have yet another character. In this way consciousness develops greater expression, the more articulated the system of self-organizing intelligence the more evolved the structure manifesting in existence.

Molecular configurations are similar to human beings, in that, a person with one character, working with another person with a different character, creates a new dynamic or relationship. Just as people form relationships between each other, atoms have their own very simple relationships with each other; both are based on how the consciousness is being expressed through them and between each other. Molecules also act differently depending on the conditions in which they exist. Water, H_2O, has one characteristic when frozen as ice, another as liquid and yet another as gas when heated. Temperature and pressure create different conditions which result in different relationships.

Molecular structures, compounds and elements in general are combinations of atoms working together to form relationships that further evolve the expression of consciousness through their association. Each system has in common attractive and repulsive forces, a field bonding the atoms, a current running through the system and a particular pattern in which they interact forming their character. Each new system further evolves the consciousness being expressed through them.

As these configurations of atoms or systems interact with other systems, under different conditions, they create increasingly more complex or articulated forms of expression. The consciousness being expressed in more articulated ways evolves the intelligence of the system. These systems evolve into larger chains, such as RNA and DNA, and serve as the building blocks of cell structures.

Cells also have a positive charge on one end, and a negative charge on the other. There is a current that runs through each cell and a bio-magnetic field that runs through the cell that defines its character. Before a cell divides, the chromosomes align by the polarity within the cell. The design or blueprint of the cell is held within its field, represented by the chemical constructs we call RNA and DNA.

Each cell has its own innate intelligence. Each takes in nutrients, eliminates waste, interacts with others in its environment, communicates or interacts with others of its kind, and can learn or adapt to conditions in its environment. It is a little being in itself. Like single celled phytoplankton, or diatoms, each is its own being expressing its consciousness, albeit in a very rudimentary way.

Cells have attractive and repulsive forces, and can work together to form increasingly more complex life forms, evolving into plants and animals. Every plant and animal has its own life force, attractive and repulsive forces within it, and maintains a field which contains its system of organizing intelligence and has its own unique character. Consciousness expresses itself uniquely through each living organism.

All living organisms work as a part of a larger system of innate intelligence which on Earth we call the biosphere or ecology. Each exchanges attractive and repulsive forces, and each forms bonds with others of its kind and is repulsed by other forces. Each learns, grows and evolves into increasingly more complex systems of organizing intelligence. In humans, we organize into families, clans, companies, cultures, societies and civilizations.

Humans, animals, plants and minerals comprise a greater system of organizing intelligence. Each is a part of a greater whole, and each draws its existence from the energy of the Earth itself. There is a positive charge on the North Pole, and a negative charge on the South Pole. There is a field around the globe, and a current that flows through and around the Earth. The innate intelligence of the Earth is held within the field surrounding the Earth.

The Earth is a part of the solar system with the Sun at its center. The Sun too has an electro-magnetic field; it too organizes the systems that orbit around its field and conducts the life force energy for life around it. Consciousness is being expressed through the Sun and the planets, and this is what provides for life on the Earth. We have no existence without this guiding energy and the consciousness guiding it.

The Sun too, like all stars, is part of yet another system of organizing intelligence we call a galaxy. The galaxy also has an electro-magnetic field. It too contains both attractive and repulsive forces and conducts the creation and destruction of stars within its field. It creates and destroys life, and conducts the cosmological conduct of all the stars within it. It too has an innate intelligence as the consciousness expresses through its manifestations of energy and matter, stars and planets, and life forms.

Each galaxy is part of yet another greater cosmological system of organizing intelligence we call a universe. The universe itself is alive, it contains attractive and repulsive forces, and it conducts and guides energy, matter and mass. It expresses the consciousness in an articulated matter that defines the characteristics of life we call natural laws or physics. It creates and destroys galaxies. The universe is an expression of the self-reflecting nature of con-sciousness; the universe is God reflecting upon Its own potential. God is dreaming universal existence.

Chapter 5 – A Physical Transformation

As the consciousness expresses throughout the universe, a spreading wave of energy manifests as sub-atomic particles, creating a dynamic force pushing outward and an attractive force pulling inward. From the inside out, consciousness expands in spinning waves and atoms push away, stars push away, galaxies push away, and the entire universe is pushing away, creating time and space. The universe is expanding. From the inside out, the entire universe is expanding in all directions. This expansive force we call the Big Bang, and the Big Bang is *still* exploding. This expansive force surfaces as the most basic element - hydrogen.

Hydrogen in its origin has one proton and no electron. The proton represents the expansive force. The entire universe is filled with hydrogen; it is the most abundant chemical element in the universe. Seventy five percent of all mass in the universe is hydrogen. It is the lightest element with an atomic weight of 1. The spinning force of the single proton in the hydrogen atom in aggregate creates a massive expansive force throughout the universe. From within the field created lies a reciprocal attractive force from the static state within it which is gravity.

Vast regions of space are filled with hydrogen atoms. These fields of gas and dust are called nebula. Most of these nebula fields are filled with hydrogen, but other heavier elements are drawn into the field by the explosions of Super Nova. Super Nova are the result of the destruction of large stars. As the collective, attractive forces of hydrogen atoms pull their atoms together, the force of this gravity joins these atoms together into a mass we call a star.

As more hydrogen is gathered together creating more mass, the pressure and heat from within the star increases creating the energy we call a star. This energy is known as nuclear fusion. Within each star there is a force pushing outward from the fusion reaction as well as inward from the gravity of the mass. This interaction of forces pushing inward and outward, and the ultimate creation and destruction of stars from these forces, create the elements of the universe – the building blocks of life.

Our galaxy alone has more than one hundred billion stars. Our universe has over one hundred billion galaxies. There are more stars in the universe than there are grains of sand upon the Earth. Galaxies are cities of stars. As the galaxy spins it creates and destroys stars, and as the universe spins it creates and destroys galaxies. The universe spins, the galaxies spin, the stars spin, the planets spin and the energy within all elements and life forms spin – this is Shiva's dance.

The Earth and Sun are part of the same expression of consciousness. We have no existence separate from the Sun. The mass of our own human body was created by the Sun, and the energy causing life within us is given via the Sun. We feed off the Sun's energy. As the Sun rotates, fields of energy within it clash creating giant magnetic loops. These forces build up and then snap creating coronal mass ejections, or solar wind. This energy fills our solar system and affects the Earths' magnetic fields and tectonics. As the Earth reacts to these coronal mass ejections the Earth's electro-magnetic pulse is disrupted and this causes movements in the Earth's crust, which, in turn, causes earthquakes, tsunamis and vulcanization, and can cause power lines and electronic devices to go dead.

As the Earth was forming, some 4.54 billion years ago, the heavier elements sank to the bottom and settled in the core, while the lighter elements rose up to the crust. Around the core molten liquid iron began to spin as the Earth spun creating essentially an electro-magnetic engine. This electro-magnetic engine is what supports life on Earth. This electro-magnetism emits a positive charge out of the North Pole, and a negative charge from the South Pole. This positive force on the North is what draws a compass needle north. The Earth's polarity also changes periodically, with north and south shifting charges every twenty thousand years or so. This electro-magnetic field covers the whole Earth and protects the Earth from solar winds or radiation. Without it, we would not have an atmosphere or life as we know it.

The electro-magnetic current imbues the atmosphere and the mass of the Earth itself with electro-magnetism. This electro-magnetic field fills the air we breathe with a subtle current of energy. When we need more energy we start breathing more, when we consume less energy our respiration naturally decreases. All life forms on the planet draw from this current of energy for life. The yogis call this vital energy "prana". This field charges our bodies with energy, and the entire body of a human being, or any living being, is filled with bio-magnetism. The Sun's energy further activates the Earth's energy, which in turn is transmitted to every human being. We are all connected energetically. The same consciousness that created and sustains the Sun created and sustains the Earth, and all life on the Earth.

The nuclear reaction from within each star releases energy. Light, heat, and electro-magnetism fill our solar system, and every solar system. Our Sun is over 4.6 billion years old, and lies 93 million miles away. It is massive compared to our Earth. A million Earths could fit inside the Sun. However, there are stars a billion times the size of our Sun. Our Sun is of average size as far as stars go. Each star is emitting energy caused by the fusion of atoms as hydrogen is being converted into helium, through nuclear reactions within the stars. One form of energy that is released from the Sun is called a photon, and photons provide the experience we call light.

Science identifies light as a particle, but a photon is actually part of a spreading wave of energy and the particle is only what appears on the surface. There is a continuous stream of energy spreading out in all directions from the Sun. Light travels at 670 million miles per second, so it takes approximately eight minutes for the light of the Sun to reach the Earth. The light from other stars has traveled from millions to billions of years ago; it's like looking back in time.

The light that fills our life is running in continuous stream from the Sun. It's not just something that was at the Sun and now is here on Earth, we are connected to the Sun through the rays touching us. Whether seen or not, the energy from the Sun, as light, heat and electro-magnetism is entering our body continuously. So is the energy from the billions of other stars and galaxies. We are all interconnected through the expression of consciousness, though this life-force energy from the cosmos. Both the energy of the Sun and the consciousness being expressed through the energy from the Sun, and other stars, is providing life on Earth in all its life forms; like the electricity that runs an appliance.

The energy of our Sun is moving out in all directions, as is the energy from all the other stars in our galaxy and other galaxies. All the stars are interfacing through fields of light and magnetism which form the time space continuum. As the light of the Sun enters into your eyes, that energy is actually taken into your body, and the photons picked up in your eyes are sending a signal into your brain. Humans are in a continuous interface with the Sun and all the stars and galaxies. We are all being directed and guided by a cosmic intelligence. The same forces that created our Sun and Earth are creating other solar systems and planets with similar elements throughout the universe. The same consciousness is pervasive and is not bound by time or space, and as a matter of fact, it's creating time and space.

Humans are made of star dust. Hydrogen and helium are by far the most abundant elements in the universe. However, iron is the most abundant element (by mass) making up the Earth, and oxygen is the most common element in the Earth's crust. The production of heavier elements, from carbon to the heaviest elements like iron, occur through what science calls stellar nucleosynthesis, and made available later in our solar system and planetary formation by supernovae. As previously mentioned, a super nova is a massive explosion that disseminates the heavier elements throughout the stellar neighborhood. The high abundance of oxygen, silicon and iron on the Earth reflects their common production in such giant stars and their explosions.

The heavier elements gathered closer to our Sun due to gravitational forces when our solar system was forming from within a massive elemental gas formation or nebula cloud, while most of the lighter elements gathered in the outer planets. Thus, Mercury, Venus, the Earth and Mars have solid mass, and the outer planets of Jupiter, Saturn, Uranus and Neptune are made of lighter elements – mostly gasses. The temperature differences and pressures created from the various distance from the Sun formed different elements and structures in each of the planets and their moons. The Earth's proximity, not too close or too far, from the Sun enabled the formation of water which supports the life forms we have today on the Earth.

While all of the 98 naturally occurring elements known to exist on Earth have been identified in mineral samples from the Earth's crust, only a small minority of elements are found in recognizable relatively pure minerals. The most common native elements on Earth are copper, silver, gold, carbon (as coal, graphite and diamonds), sulfur and mercury. While about 32 of the chemical elements occur on Earth in native uncombined form, most of these occur as mixtures. As an example, atmospheric air is primarily a mixture of nitrogen, oxygen and argon, and within the solid elements we see them as alloys such as iron and nickel.

Two dozen of the Earths elements are essential for various forms of biological life. Most organisms share common elements in their structures. There are a few differences, such as ocean algae use bromine but land plants do not and animals don't seem to need any. All animals require sodium, but some plants do not. There are just six elements that make up 99% of the mass of the human body: carbon, hydrogen, nitrogen, oxygen, calcium and phosphorus. In addition to these elements that compose the body mass, humans also require the consumption of at least a dozen other elements in the form of chemical compounds.

The Earth formed around 4.54 billion years ago by accretion from the solar nebula. In addition to the elements found in the Earth's crust, water was brought in by comets and asteroids which condensed into clouds. As the Earth cooled, this caused condensation which, in turn, formed the oceans. Some of the primary elements that occurred on the surface and early atmosphere included: water, methane, ammonia and hydrogen, as well as carbon dioxide, hydrogen sulfide and sulfur dioxide from volcanic activity.

In 1952 an experiment done at the University of Chicago by Stanley Miller and Harold Yuri, mixed methane, ammonia, hydrogen and water in a closed system of glass tubes and flasks to simulate an early Earth's atmosphere. Water was heated to induce evaporation to simulate the early Earth's atmosphere and electrodes were placed inside and fired to simulate lighting which would have been common in our primordial atmosphere. Within a week of the experiment 10-15% of the carbon created was now in the form of organic compounds. 2% of the carbon had formed amino acids which are the building blocks of proteins that form living cells.

Within this primordial soup it was discovered that sugars and lipids were also found; thus the basic building blocks of life were manufactured by basic natural life processes. Similar experiments mixing hydrogen and cyanide, ammonia and water in laboratories have resulted in chemical reactions forming numerous amino acids which are the workhorse molecules of life. These amino acids are used in everything from structures that form cells, and create tissues like hair, skin and nails, as well as enzymes, the catalysts that speed up or regulate chemical reactions. These same processes are replicated throughout the universe.

In March of 2009, researchers at NASA's Goddard Space Flight Center reported the discovery of the amino acid Isovaline in samples of meteorites that came from carbon-rich asteroids. Dr. Michael Callahan's research team at NASA has also indicated that the building blocks for DNA have also been found in asteroids and comets. The team found adenine and guanine, which are components of DNA called nucleobases, as well as Hypoxanthine and Xanthine which are used in other biological processes. In two meteorites, the team discovered for the first time trace amounts of three molecules related to nucleobases: purine, 2,6-diaminopurine, and, 6,8-diaminopurine. These compounds have the same core molecule as nucleobases but with a structure added or removed. Thus, the basic building blocks of life are routinely manufactured in space.

In 1984 meteorite hunters from the ANSMET project discovered a four pound meteorite in Allan Hills, Antarctica, that was originally from Mars during a time when Mars had an atmosphere and water. The meteorite apparently blew off the planet due to a large asteroid impact on Mars. In 1996, under the scanning of an electron microscope, evidence of fos-silized bacteria-like-life forms was discovered within the structure. This carbon laden meteorite known as ALH 84001 was full of amino acids and carbonate globules that are the remains of nanobacteria which were apparently living on Mars some 3.6 billion years ago. In addition to various amino acids, polycyclic aromatic hydrocarbons (PAH) were also discovered. These PAH's also indicate that microscopic life forms existed on Mars.

Consciousness expressing in a multiplicity of combinations creates the variety of environments and conditions for various forms of life to evolve in similar ways. Water, carbon and other elements needed to support life are abundant in the universe, and so is the radiation that infuses the energy that causes the chemical reactions needed to create the building blocks of life forms.

Thus, life naturally evolves from the interactions of elements and molecules under conditions created from the stars, and distributed to planets, comets and asteroids. Consciousness evolves its expression through the formation of self-organizing systems creating ever greater forms of intelligence. We see the basic patterns for life being replicated throughout the universe, and can observe its evolution here on the Earth.

Chapter 6 – Self Organization of Life

Atoms are systems of energy interacting in ways that form patterns. Each atom has within it both an attractive and a repulsive force that causes it to either be attracted to other atoms and elements or be repulsed by them. The places where atoms come together and bond are chemical substances, and they can be solids, liquids, gases or plasma depending on the environment in which they exist. As these systems form more complex structures and the energy flowing through them becomes more articulated, they evolve into more advanced systems for expressing their inherent consciousness. The configurations of articulated energy patterns within and between atoms forms the underlying basis for more advanced systems. The expression of consciousness being articulated in increasingly more complex ways forms the basis of intelligence.

An element is a chemical substance that is made up of a particular kind of atom and hence cannot be broken down or transformed by a chemical reaction into a different element, though they can sometimes be transmuted into another element through a nuclear reaction. There are about a 120 known elements, about 80 of which are stable, which means they do not change by radioactive decay into other elements. However, some elements can occur as more than a single chemical substance and these are called allotropes. For instance, oxygen can exist as both diatomic oxygen (O_2) and as Ozone (O_3). Elements can also bond into combinations forming other substances through chemical reactions, and these are called chemical compounds.

Chemical compounds can either bond atoms together in a molecule or form crystals. In other words, the atoms begin building three dimensional structures, which further articulates the way the energy within them interacts. This sacred geometry or architecture further develops the character of the atomic structure into increasingly more advanced systems of organizing intelligence. The attractive forces of atoms, molecules and ions form crystalline lattices. These structures provide for a greater expression of the consciousness and how the energy moving through these structures can interact within themselves and with other atomic or chemical structures.

One of the primary building blocks for chemical structures is carbon. Carbon is a very large and adaptable atom that can be used to develop multi-tiered and sided structures. So compounds based primarily on carbon and hydrogen atoms provide for the complexities of organic life. Thus, molecules made of carbon and hydrogen are called organic compounds, and all others are called inorganic. So science calls those more complex structures that can support more sophisticated expressions of consciousness "carbon based life forms".

The more ways that atoms can interact and build upon one another, the more sophisticated the system of organizing intelligence is. These more complex systems channel the energy flowing through them in more articulated ways. Systems building upon other systems, and working with other systems, evolves the consciousness being expressed in increasingly more complicated ways and provides more options and interactions with other atoms, molecules and chemicals.

The development of these systems provides for the basis of intelligence. By controlling information flow through biochemical signaling and the flow of chemical energy through metabolism, biochemical processes give rise to the complexity of life.

You can look at the chemical exchange of energy between molecules like the exchange of words in children. In the beginning, the exchange of needs and intentions is very basic. But as the child acquires new words to express new needs and desires, the language of the child develops. Chains of words form more sophisticated language, and clearer and more defined expression of needs and desires develops. Similarly, as organic systems of energy develop, such as molecules, they communicate through exchanges of energy signals. These signals are expressed through chemical or electrical impulses that express or communicate their needs and desires. In each system, the communication becomes more articulated as they develop creating a greater expression of the consciousness desiring to be communicated.

Through this evolving biochemistry more developed molecular structures, functions and interactions build cellular components such as proteins, carbohydrates, lipids, nucleic acids and other biomolecules. One of the primary building blocks of life is called amino acids. An amino acid is a combination of carbon, hydrogen, oxygen and nitrogen atoms. Amino acids serve as the building blocks of proteins, which are linear chains of amino acids. Amino acids can be linked together in varying sequences to form a vast variety of proteins. Amino acids have polar and nonpolar fields, creating places where they can adhere to or be repulsed by other atoms. Like the toy *LEGOS*, building blocks with varying numbers of pegs, they fit compatibly in some combinations but not in all.

Amino acids also build non-linear systems, creating what are called "branched-chain amino acids". These enable the architecture of the amino acids to branch out in different directions, creating even more possibilities for systems to evolve. As the amino acids form increasingly more complex systems, they form short polymer chains called peptides or longer chains called polypeptides or proteins. The process of making proteins is called translation and involves the step-by-step addition of amino acids to a growing protein chain as directed by DNA.

Like building multiple rooms for different purposes, the amino acids provide for various molecular utility. The activity that occurs in each part is like people holding different jobs in a company, family or society. As more proteins form, they can perform more specialized functions and add a depth of character to the molecular body being created. They can perform multiple functions and act in many more ways. They can also communicate in more sophisticated means and they develop organizational structure. In this way, the consciousness being expressed through them develops, just as humans developed from simple tribes to societies and civilizations. Some proteins are construction workers, some clean house, some gather food, and some direct all the activity going on. They all work together as a team, within a collective consciousness.

Some proteins are enzymes and can catalyze biochemical reactions that are vital for metabolism. Other proteins play a role in structural or mechanical functions, such as creating actin and myosin used in building muscle or skeleton structures, and others serve as a form of scaffolding that helps maintain cell shape. Other proteins are used in cell signaling, immune responses, cell adhesion and the cell cycle. Each protein has its own unique amino acid sequence that is specified by the nucleotide sequence of the gene encoding the protein that $\Delta 90$ come from the DNA or RNA. The DNA serves as the brains of the organism.

Proteins play an important part in cell signaling and signal transduction. Molecules and cells communicate with each other energetically and chemically. Some proteins, such as insulin, are extracellular proteins that transmit a signal from the cell in which they were synthesized to other cells in distant tissues. Others are membrane proteins that act as receptors whose main function is to bind a signaling molecule and induce a biochemical response in the cell. Thus proteins help evolve intelligence via the ability to communicate with others of their kind. This is yet another tier in the expansion of consciousness that evolves its expression through intelligence.

There are also many receptor sites on the surface of cells and an effector domain within the cell, which may have enzymatic activity or may undergo a conformational change detected by other proteins within the cell. Cells pick up the field energy around them and conduct the information being received, which creates a change in behavior in the cell. Antibodies are protein components of an adaptive immune system whose main function is to bind antigens, or foreign substances in the body, and target them for destruction. Thus, cells communicate with one another and can adapt to changes in their environment. In other words, they learn, and grow, and evolve a rudimentary intelligence.

Other biomolecules important to mention are carbohydrates. Carbohydrates are made from monomers called monosaccharides, which are basically sugars or food, such as glucose, fructose and deoxyribose. Carbohydrates are important for their ability to store energy and genetic information, and play an important role in cell to cell interaction and communication. Lipids are another essential molecule that are essentially various combinations of fatty acids, and these can form molecular structures that give rise to form.

Nucleic Acids are an essential component for the development of life forms. Nucleic acids are the molecules that make up DNA and RNA, which store genetic information needed for cell function and replication. As the cells receive feedback from their environment through electrical and chemical interactions in their environment, that data is imprinted in the fields of awareness within the molecule and stored within the field. This self-reflective nature within the consciousness of the molecules creates the patterns which form the coding within the DNA. These patterns can then be duplicated or read by other parts of the molecule.

RNA stands for ribonucleic acid and DNA stands for deoxyribonucleic acid. DNA consists of two long polymers of simple units called nucleotides, with "backbones" made of sugars and phosphate groups joined by chemical bonds. It is the sequence of these four nucleobases along the backbone that encodes information. In other words, these patterns take the form of molecular structures, like a chemical alphabet. This information is read using a genetic code, like molecular letters within a chemical alphabet. The code is read by copying stretches of DNA into the related nucleic acid RNA in a process called transcription.

Within cells, DNA is organized into long structures called chromosomes. You can think of chromosomes as a chemical instruction manual for a life form. During cell division these chromosomes are duplicated in the process of DNA replication, providing each cell its own complete set of chromosomes. Genes correspond to regions within the DNA, a molecule composed of a chain of four different types of nucleotides. It is through these sequences of nucleotides that genetic information is passed on. Genetic imprints are made and stored via these nucleotides; like taking notes or encoding a program through a sequence of molecular characters.

The four nucleotides are: adenine (A), cytosine (C), guanine (G) and thymine (T). Information about cell shape and activity is encoded in these four molecules, like the zeros and ones used in computer code. The specific sequence identifies specific characteristics about the cell composition and behavior. As energy is passed through the code, it reads these encoded symbols like a blueprint on how to build cell structure and how the cell is to function. DNA is genetic programming. The self-reflective quality of consciousness provides feedback to the cell about which characteristics and traits work for the organism and which do not and need to be adapted.

Each nucleotide in DNA preferentially pairs with its partner nucleotide in a twisting pair of strings forming a double helix. In this sequencing, "A" pairs with "T", and "C" pairs with "G." In its two stranded form, each strand effectively contains all necessary information, redundant with its partner strand. These DNA strands are often extremely long; the largest human chromosome, for example, is about 247 million base pairs long. Herein lie the genetic makeup of all life forms which determine, to a large degree, what physical features, characteristics and behaviors a human being or other life form will inherit. Thus, human programming is passed on through genetic lines.

RNA then serves as a messenger that delivers the DNA code through a ribosome, which reads the RNA sequence and copies it. For a cell to divide, it must replicate the DNA in its genome so that the daughter cells have the same genetic information as their parent. Breaks or adaptations in the sequencing of this code are the basis of mutations that result in changes in the design and function of life forms. In other words, molecular patterns can be cut and pasted into new sequences. Patterns or processes that don't work can be erased, and new patterns can be adapted or modified. These adaptations are what enable life forms to adapt to changes in their environment.

All organisms have many genes corresponding to various biological traits, such as eye color, hair type and color, or number of limbs. The genetic code is the set of rules by which a gene is translated into a functioning protein. The energy that moves from one end of the sequence to the other forms the pattern that directs the manufacturing of molecules to create the desired cells. Thus, an innate intelligence in the body of the cell is directing how cells are made, and by directing how cells are made this innate intelligence directs how all life forms are made.

As these systems of organizing intelligence start working together they form cohesive units called organelles. An organelle is a specialized subunit within a cell that has a specific function and is usually separately enclosed within its own lipid bilayer. The name organelle come from the idea that these structures are to cells what organs are to the body. Some of these organelles include the chloroplast which traps energy from sunlight to create photosynthesis. This provides energy for the cell to exist. Some proteins that are made in the cytoplasm contain structural features that target them for transport into the mitochondria or the nucleus, thus providing feedback to the cell.

The endoplastic reticulum is used for the translation and folding of new proteins, similar to how an organism regenerates cells. Rough endoplasmic reticula are covered with ribosomes for this conversion; the ribosomes read the molecular signals. The golgi apparatus sorts and modifies proteins for specialized uses. It works something like a post office. It receives items (proteins), packages and labels them, and then sends them on to their respective departments within the cell. From the golgi, membrane proteins can move to the plasma membrane, to other sub-cellular compartments, or they can be secreted from the cell. The endoplasmic reticulum and golgi can be thought of as the "membrane protein processing compartment."

The mitochondrion is used for energy production from the oxidation of glucose substances, or sugar. The vacuole stores key chemicals and helps maintain homeostasis. The nucleus is where DNA is stored and maintained; it controls all the activities of the cell, and is involved with RNA transcription. It's like the cells' brain. So a cell has a way to determine its needs and act upon those needs similar to our own brain; it take in nutrients and digests them and eliminates waste; it builds its own tissues or membrane; it regulates it cell functions; and it also reproduces. Cells also communicate with others of their kind. They are able to identify things they need as well as avoid those things they don't want or what might hurt them. It learns, it develops, it adapts to changes in its environment, and it communicates its desires. Its consciousness is self-reflective.

Some cells act alone, others in groups or clusters, while others form whole colonies. These groups of cells working together are called micro-organisms and were discovered in 1675 by Anton van Leeuwenhoek while using a microscope of his own design. Micro-organisms include bacteria, fungi, achaea and protists; microscopic plants such as green algae; and small animals such as phytoplankton and diatoms.

Thus, groups of cells create more advance life forms, which enable consciousness to evolve to higher levels of intelligence.

Viruses are smaller and simpler organisms that typically rely on a host cell for their survival. Although they have genes, they do not have a cellular structure, which is typically how science views the basic unit of life. Viruses do not have their own metabolism, and require a host cell to make new products. They reproduce by making multiple copies of themselves through self-assembly. There are millions of different types of viruses. They are found in almost every ecosystem on Earth and are the most abundant type of biological entity on the planet. In evolution, viruses have played an important role in horizontal gene transfer, which increases genetic diversity. They come in a wide variety of shapes and sizes, typically specialize in the type of host cells in which they reside, and can be transferred through almost every kind of interaction between various life forms.

Single-celled microorganisms were the first forms of life to develop on Earth, approximately 3 to 4 billion years ago, and for the first 3 billion years of life on Earth these microorganisms remained microscopic. Bacteria has been identified in fossils dating back

3.5 billion years, and DNA samples of living bacteria have been discovered in the permafrost of Canada, Siberia and Antarctica dating over half a million years. Microorganisms can freely exchange genes by conjugation, transformation and transduction between widely divergent species. This horizontal gene transfer, coupled with a high mutation rate, enables microorganisms to adapt to changes in their environment (genetic variation) and allows them to swiftly evolve to survive environmental stresses.

Bacteria and archaea are the most diverse and abundant group of organisms on Earth and they inhabit almost all environments where some liquid water is available. They are found in sea water, fresh water, soil, air, within animal's gastrointestinal tracts, in hot springs and even deep beneath the Earth's surface. The number of these microorganisms is estimated to be around five million trillion trillon, and account for at least half of the biomass on Earth.

Some of these microorganisms have adapted to extremely hostile environments on Earth like the poles, deserts, geysers, rocks, within the depths of the ocean and near volcanic vents and all these that live under extreme conditions are known as Extremophiles. Extremophiles have been found thriving at temperatures as high as

130 C (266 F) and as low as -17 C (1 F). They are also found in areas with an acidity/alkalinity of less than pH 0 to up to pH 11.5; in salinity up to saturation; pressure up to 2,000 atm, and down to 0 atm (the vacuum of space); and radiation levels up to 5kGy. This indicates that microorganisms can exit in a variety of environments outside the planet Earth, and evidence suggests that they do.

Bacteria are one of the most diverse and abundant life forms on the planet. There are typically 40 million bacterial cells in a gram of soil and a million bacterial cells in a milliliter of fresh water. Their biomass exceeds that of all plants and animals upon the Earth. There are approximately ten times as many bacterial cells in the human flora as there are human cells in the body, with most residing in the skin or intestinal track. Some forms of bacteria are necessary to sustain human life, while others can be deadly.

The bacterial cell is surrounded by a lipid membrane, or cell membrane, which encloses the cell and acts as a barrier to keep nutrients, proteins and other components within the cell, as well as helping to keep other matter out. In many forms of photosynthetic bacteria, the plasma membrane is highly folded with a light-gathering membrane to collect energy. They come in a wide variety of shapes and sizes, and sometimes have a tail that helps them swim or wiggle their way through their environment.

Bacteria often function as multicellular aggregates known as biofilms, with the ability to exchange molecular signals for inter-cell communication and they engage in coordinated multicellular behavior. They coordinate and work together as a team. This effort enables the division of labor, accessing resources that cannot be effectively utilized by single cells, and they can collectively defend themselves against antagonists. Some types of bacteria even have the ability to regulate their population control, what scientists call quorum sensing. In other words, they form communities and work together as a team with a collective consciousness; they have innate intelligence.

The next evolution in microorganisms is algae. Algae range from unicellular to multicellular forms, and range in size from microscopic to giant kelp that grow to 65 meters in length. Algae are photosynthetic like plants, but simpler in that their tissues are not organized into distinct organs found in land plants. They are the precursors to land plants. Algae are found in the fossil record dating back to approximately three billion years in the Precambrian period.

Some forms of algae form symbiotic relationships with other organisms, where the algae supply photosynthates or organic substances to the host organism that in turn provide protection to the algae cells, such as in the cases of lichens, coral and sponges. There are believed to be tens of thousands of varieties of algae spread out around the world.

Diatoms are a major group of algae and one of the most common types of phytoplankton. Most diatoms are unicellular, although they can also colonize in the shape of filaments or ribbons. Phytoplankton are photosynthesizing microscopic organisms that inhabit the upper sunlit layer of almost all oceans and bodies of fresh water. Phytoplanton are the primary agents for the creation of organic compounds from carbon dioxide dissolved in water and lie at the heart of the food chain for our oceans.

Phytoplankton accounts for half of all photosynthetic activity on the Earth; they are responsible for much of the oxygen present in the Earth's atmosphere – half of the total amount produced by all plant life on the planet. They were the primary agents for creating the air in the atmosphere that most life forms need to survive. Protozoon are yet another type of unicellular organism, diverse in nature and are similar to algae in composition and exist all over the world.

Fungus is a type of microscopic organism that also develops multi-cell structures. Fungi include yeasts, molds and mushrooms. They are different in cell structure than bacteria, plants and animals, but genetic studies indicate that fungi are more closely related to animals than plants. Fungi perform an essential role in the decomposition of organic matter and have a fundamental role in nutrient cycling and exchange. Some fungi like mushrooms and truffles can be eaten, while others are deadly poisonous, and yet others provide hallucinogenic properties that have played an integral role in the spiritual development of indigenous cultures around the world, while molds and yeasts have been used as leavening agents for bread, and in fermentation processes in wine, beer and soy sauce.

There are over 1.5 million species of fungi, but only around 5% have been formally classified. They range from single-celled aquatic chytrids to large mushrooms. Fungal cells contain membrane-bound nuclei with chromosomes that contain DNA. Fungus serves as another example of how innate intelligence developed a means for single cells to work together to form a more complex collective multi-celled organism as it evolved.

So, from chemical elements and compounds that developed systems of self-organizing intelligence, the fields of consciousness working within these units expanded and created relationships that served the interests of the whole to survive. Organelles developed a means of transmitting information that enabled all the parts to work together as a whole, forming cells. The awareness of cells expanded into groups of cells that created even more complex systems of self-organizing intelligence that became simple life forms. At each stage in evolution the consciousness was expanded into ever greater systems of self-organization, and as a result a higher level of intelligence developed.

Chapter 7 – The Evolution of Consciousness

Evolution is the evidence of consciousness actively guiding the development of life. God's invisible hand becomes visible through the inner workings of creation. Just as atomic interaction developed systems of self-organizing intelligence on Earth, similar cosmic, geologic and biological developments have been and are taking place throughout the galaxy and universe. Not only is consciousness evolving on the physical plane of time, space, energy and matter, but on other planes not received through the physical senses or mental cognition. However, all systems of organizing intelligence are working together as one functioning whole unit.

We can observe evolution from the smallest systems of sub-atomic particles, atoms, compounds, molecules and cells, to living organisms, biospheres, solar systems, galaxies and universes. There are systems within systems all interconnected and functioning as one unit of expression of consciousness on the physical plane, within multiple planes operating within the whole of existence. Life itself is a reflection of the consciousness in a multiplicity of refractions upon the surface of its own nature – life is God reflecting upon the possibilities of Its own existence and manifesting it.

Life evolves through the creation of life forms, and we typically call this expression of life Spirit. Spirit is consciousness being expressed, and as it is being expressed through various interactions and associations in its environment it naturally evolves; this is most easily recognized in human beings. The human mind is a reflection of this innate process for consciousness to evolve into ever greater degrees of self-expression, just as humans themselves are evolving to greater degrees of self-expression. Human existence is the evidence of consciousness in expression. As we evolve, we are becoming more conscious of our own nature, and of nature itself.

Evolution is defined as any change across successive generations in the heritable characteristics of biological populations. Evolution is what gives rise to the diversity of life at every level of biological organization, including species and individual organisms and molecules such as DNA and proteins. On the Earth, organic life evolved from a universal common ancestor approximately four billion years ago. Through repeated speciation, the divergence of life can be inferred from the shared common sets of biochemical and morphological traits, indicated by shared DNA sequences. We can reconstruct this evolution by tracing the genetic codes within all life forms and observing the fossil record which confirms these patterns of evolution.

The idea that one animal could descend from another type, or evolve from a more primitive organism to one more complex existed within the ancient theories of the yogis and the concept of karma. The first Western descriptions date back to the pre-Socratic Greek philosophers, such as Anaximander and Empedocles. Later Aristotle explained that all natural things were actualizations of different fixed natural possibilities known as "forms", "ideas", or "species". He observed that all things have an intended role to play in a divine cosmic order.

In 1751, the French naturalist Maupertuis observed that the natural modifications occurring during reproduction and accumulating over many generations produce new species. However, this thinking was subdued by conventional religious dogma that misunderstood the meaning of the Biblical metaphor in *Genesis* of Creation being over a period of Earth days, where the correct meaning in Judaism is metaphorical and the interpretation of "days" actually refers to epochs of time and not physical Earth days. The interval of time we call a day has been changing, and continues to change, as the Earth's rotation continues to slow down.

The naturalist Georges-Louis Leclerc the Comte de Buffon (1707-1788), later suggested that species could degenerate into different organisms, and Erasmus Darwin (1731-1802) proposed that all warm-blooded animals could have descended from a single micro-organism or "filament". Lamarck's "transmutation" theory of 1809 envisioned simple life forms developing greater complexity in parallel lineages with inherent progressive tendencies, but it was Charles Darwin in 1859 who published *On the Origin of the Species that explained* how natural selection led to the evolution of life forms, including human beings.

Darwin suggested that evolution by natural selection is a process that can be inferred by three facts about populations: 1) more offspring are produced than can possibly survive, 2) traits vary among individual species, leading to different rates of survival and reproduction, and 3) trait differences are heritable. Thus the populations that survive are the ones that can adapt to changes in their environment and that these changes are passed down through genetics, as well as through epigenetic (psychosocial) and external environmental conditions. Heritable traits are known to be passed on from one generation to the next via the species DNA, as discussed in the preceding chapter.

Descendants inherit not only genes but environmental conditioning generated by ecological actions of ancestors and learned behavioral responses passed on through individual and collective learning. The self-reflective nature of consciousness enables organisms with traits that give them an advantage over their competitors to pass on these advantageous traits to their heirs, while traits that do not confer an advantage are not passed on to the next generation. It's the survival of the fittest that provides the development of evolution to life forms that have a more competitive advantage.

Evolution influences every aspect of the form and behavior of organisms. These adaptations increase fitness by developing the ability to find food, avoiding predators and attracting mates. Organisms can also evolve by cooperating with each other and developing symbiotic relationships. There are macroevolutionary developments that occur at or above the level of the species such as extinction and speciation and micro-evolutionary developments such as adaptation which occur within a species. Adaptation enables organisms to make organic changes within themselves to adapt to changes in their environment. We see evidence of this in the wings of flightless birds, hip bones in whales and snakes, and vestigial structures in human beings like wisdom teeth and the coccyx or tail bone.

Examples of co-operative adaptation include the relationship between plants and mycorrhizal fungi that grow on their roots and aid the plant in absorbing nutrients from the soil or in the bacteria within human intestines that help it digest its food. Another example are insects and birds serving as an essential component in the dissemination of pollen and seeds for flowers and plants to propagate, which also provides food for animals disseminating the pollen or seed. They work together compatibly. There is a greater system of organizing intelligence directing the interaction between and within the species.

Extinction also plays a crucial role in the evolution of consciousness. Those species that don't contribute or learn to adapt to the changes in their environment become extinct. Mass extinctions are more common than adaptation. Over 96% of all know species on Earth have become extinct. Our present day extinction rates are from 100 to 1000 times greater than those of the past, and up to 30% of all organisms on the Earth may become extinct within the next century.

Soon after the emergence of the first multicellular organisms, a remarkable amount of biological diversity appeared some ten million years ago when oxygen in the atmosphere developed as a result of the ability of microorganisms to photosynthesizes, but this was the result of billions of years of adaptations from microorganisms and the environment. If you follow the timeline of evolution and the history of life on Earth, you can easy see not only the progression of evolution but that it is increasing at an increasing rate.

The Earth itself is some 4.6 billion years old. 3.8 billion years ago simple celled creatures called prokaryotes developed. 3.4 billion years ago, stromatolites developed with their ability to photosynthesize or capture the energy of sunlight and convert it into food. The byproduct of this activity created the oxygen in our atmosphere which enabled land-based life forms to develop billions of years later.

Some two billion years ago more complex cell forms developed called eukaryotes, and one billion years ago multicellular life forms began to develop. Approximately 600 million years ago very simple animals developed, and by 570 million years arthropods, the ancestors of insects, arachnids and crustaceans developed. 550 million years ago more complex animals developed, and by 500 million years ago simple fish and proto-amphibians developed. This ability to breathe air enabled various species to begin to adapt to land environments.

Approximately 475 million years ago land plants developed and by 400 million years ago insects and seeds developed creating an early ecosystem. By 300 million years ago reptiles developed, 200 million years ago mammals, and by 150 million years, birds. At 130 million years ago plants developed flowers, and by 65 million years the non-avian dinosaurs died out, changing the ecosystem dramatically and making it possible for mammals to further develop and gain greater dominion of the Earth's ecology.

Approximately 6.5 million years ago apes, began forming more complex social structures and developing the ability to walk upright. By 2.5 million years ago they began developing the most basic tools and adapting to changes in climate and the environment. By 400,000 years ago, ape-men began controlling fire and developing clans, and organized hunting and the specialization of labor. Approximately 300,000 years ago Neanderthal man developed, as well as other pseudo-human life forms in various continents around the planet. Modern Homo sapiens, from which modern humans evolved, developed around 200,000 years ago.

In hominids the fossil record shows the progressive straightening of the spine, an increase in brain volume, and changes in facial features towards becoming more refined. There is a reduction in muscles in the jaw for mastication as humanoids evolved from eating plants to eating more animals as the need for more energy arose which enabled brain development. The tail bone becomes incorporated into the pelvis as the sacrum in higher primates. Humans still have ear muscles, which were needed when as apes we moved our ears to listen to predators. Humans are born with the Jacobson's organ, which detects pheromones to trigger sexual desire, set off alarms about predators, or indicate information about food, but in early child development its abilities dwindle to the point of being useless.

In 2010 the sequencing of the Neanderthal genome indicated that Neanderthals interbred with anatomical humans around 80,000 to 40,000 years ago, around the time humans migrated out of Africa, but before they dispersed into Europe, Asia and elsewhere. Nearly all modern non-African humans have between 1% to 4% of their DNA derived from Neanderthal DNA. The competition of Homo Sapiens for land and food probably contributed to the extinction of the Neanderthal man.

Approximately 70,000 years ago, the Toba super volcano erupted in Sumatra, Indonesia, and radically changed the Earth's climate, killing most humans then alive. This may well have led to a volcanic winter with a worldwide decline in temperatures and the occlusion of the sunlight destroying the plants and animals that depend on them. Around this time we see a mass migration of Homo sapiens out of Africa and into the Middle East, and eventually over successive generations to India, Malaysia, Australia, China and Japan, and then up to Siberia and eventually to North, Central and South America. This migration has been scientifically tracked by taking samples of human DNA from around the world.

By 50,000 years ago, humans began developing basic language, clothing and fish hooks, and by 40,000 years ago, we see evidence of art as cave drawings, sculptures and even a bracelet. By 30,000 years ago, ancient man was herding reindeer and domesticating animals. They developed harpoons, needles for sewing and saws. They also began weaving baskets and cloth for clothing. By 25,000 years ago, humans were building basic stone huts and creating settlements and developed rituals such as burials and apparently worshipping a deity or deities. The earliest amongst them was apparently a fertility goddess or Venus.

By 10,000 years ago, the formations of civilizations began, and humans reached a new level of consciousness as they began to work together and further developed specialized labor, organizational structures and language. On the span of evolution, using the model of a clock, if life developed over the course of one hour human evolution was only within the last minute, and modern civilization is only the last few seconds. Here we now lie on the threshold of a whole new understanding of our self and the role we play in existence.

Humans are just awakening to the realization of their own nature and purpose; their own consciousness and potential is only now being realized.

Chapter 8 – The Evolution of Civilization

Just as atoms and molecules change form and behavior with modifications to their environments and conditions, and cells learn to adapt and interact in more complex and sophisticated ways, so do human beings. The same consciousness is expressing through all life forms, and evolving through each level of self-expression. Humans are becoming more conscious as they learn to adapt to changes in their environment, learn to work together in systems, and reflect on the nature of their existence.

As human beings evolved, they became more conscious of a higher power we call God. They began to reflect on the nature of existence and their role in existence, and they began to reflect on the nature of themselves. This self-reflection manifested in religions, philosophy, science and communication through art and literature. The development of these systems of organization and thought we collectively call civilization. Civilization marks a quantum jump in the evolution of consciousness within mankind, separating our species from other animals.

Evidence exists that cities existed before the glaciers melted and raised ocean levels well over ten thousand years ago. Over two hundred ruins have been discovered under the Mediterranean Sea. Two cities have been discovered some 120 feet under the ocean off the northwest coast of India. One is two miles wide and five miles long in the Bay of Cambay, off the State of Gujarat in Northern India, and another two hundred miles north believed to be the fabled city of Dwarka. Another possible site with large structures has been discovered underwater off an island southwest of Okinawa, and others in the Bahamas and the west coast of Cuba. Then there is also the recently discovered temple at Gobekli Tepe, in Southern Turkey, which is dated to be at least 11,000 years old. However, scientists generally attribute civilization as we know it to have begun some ten thousand years ago along the great rivers during the Agricultural Revolution. With an annual fresh water supply, people were able to grow crops year round, and this provided the ability for humans to accumulate surplus food and to grow in numbers and specialize in labor.

Arguably the oldest of these civilizations was Sumer. The Sumerians were a group of people speaking Sumerian who settled along the Euphrates and Tigris rivers within the alluvial plain called Mesopotamia, now modern Iraq. Successive settlements date back through oral tradition to over 10,000 years ago. The first settlements date back to the Eridu period in the sixth millennium BCE. This is the civilization that the Biblical Abraham was born into, and from where the stories of the great flood and history of nations are derived.

The Sumerian people who settled here farmed the land, which was made fertile by silt deposits caused by the annual flooding of the Tigris and Euphrates rivers. They were among the first to farm year round, as opposed to wandering as hunters and gatherers as their predecessors did. They are the first to have created a written language of pre-cuneiform script on clay tablets that date to 3,500 BCE. Their civilization was organized into city-states, and they were among the first to create armies and build boats and horse-drawn chariots.

The Sumerians are credited with inventing the wheel, originally a potter's wheel for making ceramics, and later for vehicles and mills. They were also among the first astrologers, mapping the constellations and giving names to these cosmological groupings in our zodiac, as well as the five visible planets. They were also the first to develop numbering systems and basic arithmetic, use messaging, and develop schools and a legal system.

The oldest story ever recorded comes from the Sumerians called the *Epic of Gilgamesh*, which is preserved on twelve clay tablets from the seventh century BCE by the Assyrian king Ashurbanipal. This predates any Biblical record by many centuries, and would have been known to Abraham who was born in the capital of Sumer, the city of Ur. In the *Epic of Gilgamesh*, the story is told about a great flood that destroys their civilization but not the whole Earth, and of a great ship that saves them and their live-stock, of a dove being sent out and returning indicating where land was; as well as the greater story of man's redemption and quest to find God and his purpose in life. Through this great event, mankind learns to renew the statutes and sacraments to God for the welfare of all people.

Another great civilization less known in the West is the Harrappan culture and the Indus Valley Civilization that developed along the Indus River in modern Pakistan and India around the same time. Seals have been found in Sumer with Indus Valley script, indicating trade and an exchange of culture and technology between these two great civilizations. They also worshiped the same primordial god of the heavens as the Sumerians, the creator god An.

An is pronounced Al as in Al-lah in Arabic and El in Canaanite, as in El Shaddai as referenced in *Genesis* in the *Bible*. This same first God is later given the name Anu by succeeding Mesopotamian cultures and in India, and later El becomes Yah or Yahweh by Moses' time and Siva or Shiva in the Hindu culture. He is also known as Ani, as in Horus Ani, in ancient Egypt. An also appears in the oldest of Hindu literature known as the Rig Veda, which in its written form dates to over 1,700 BCE. Both the East and West worship the same God. The "ah" is the common root, as well as the belief in a primordial one Creator of the Earth and Heavens.

The Indus Valley Civilization also had created cities, organized labor, cultivation and extensive trade, a written language similar to Sumerian, education, two story buildings, boats and running water. Art was well developed in the Harrappan culture, with figures showing both movement and emotions. It is here we see images of yogis sitting cross-legged in meditation. Here God was realized through an inward turned consciousness and self-reflection.

The third major civilization that developed around the same time was Egypt. The history of Egypt is the longest continuous history of any country in the world. In the Upper Paleolithic region of Egypt, ancient tools and technologies were discovered that date back to 30,000 BCE. Around 3,000 BCE, Egypt underwent a process of political unification and also created a number of city-states along the Nile River. Their culture was also based on agriculture along the Nile, and they too developed similar technologies and a written script known as hieroglyphs. Here some of the earliest spiritual texts known, such as the *Pyramid Texts*, written around 2,400 BCE, provide a code of conduct that include all ten of the Biblical commandments given to Moses, who would have been familiar with these texts having been raised in the court of the Pharaoh.

Another common theme that exists not only among the ancient Mesopotamian, Hindu and Egyptian cultures, but among almost all indigenous cultures around the world (e.g. Chinese Han, African Dogon and Zulu, European Greek and Roman, American Zuni, Mayan and Aztec, etc.), is that their civilization and knowledge of how to build it came from human-like beings that came from space. Angels, star people, gods, some extraterrestrial beings that worked for God out in space somewhere came to Earth to teach humans how to be civilized. This concept is prolific in the art, mythology and literature around the world. Mankind suddenly emerges from nomadic tribes to building cities with the help of extraterrestrial beings according to the histories of the world, including the *Bible*.

From millions of years of tribal nomadic living, all of a sudden monolithic buildings appear as well as technologies far more advanced to anything before, and the development of civilizations abruptly occur and they all attribute their own development to these celestial beings. The historical record of extraterrestrial visitations, from prehistoric cave drawings to current science, is prolific around the world and continues even to this day with over twenty governments admitting contact with intelligent extraterrestrial life. Astronauts, senior intelligence and defense officers and world leaders have come forward to attest to the proof of contact with extraterrestrial intelligence, despite its repression in the press. Extraterrestrial guidance is an integral part of human history and plays a key role in the evolution of consciousness within our species.

During this same period civilizations were also emerging along Yangtze and Yellow Rivers of China and in central Asia, but we have less archeological documentation to describe their cultures and civilizations. The rest of Asia, Europe and the Americas appeared to still be living more nomadically. As civilizations grew, so did their populations, their technology and their power. One of the common characteristics of early civilizations was the development of social class systems, and control of the land, natural resources and of other people. Slavery was used as a cheap form of labor, and the control of land and sea trade routes as a means of acquiring more wealth and power.

As a result of these clashes between expanding civilizations and the uprising of the people within civilizations, civilizations would rise and fall. As civilizations grew, their leaders created empires which sought to dominate other cultures and civilizations, but it also often brought out the dissemination of culture, technology, laws and structures that also often benefited other societies. Among the first empires to develop was Egypt in 3,100 BCE, unifying Lower and Upper Egypt.

In the 24th century BCE the Akkadian Empire (centered near modern Syria) spread into Mesopotamia, subduing the Sumerian civilization and adopting much of its culture and tradition. In 2200 BCE the Xia Dynasty arose and conquered most of China.

Over the following millennia other civilizations rose and fell. Trade became the source of power between nations and cultures. Whoever controlled the trade routes controlled the economy became the most rich and powerful. In 2500 BCE the Kingdom of Kerma developed in Sudan which controlled the exchange of goods between Africa, the Nile and the Mediterranean. The Hittites from modern Turkey controlled the routes from Central Asia to the Mediterranean around 1600 BCE, and later the Mycenaean Greeks began to dominate trade in the Mediterranean. The Persians controlled the central Asian trade routes through, first, the Median Empire (625-550 BCE), then the Achaemenid Empire (559-330 BCE), followed by The Parthian Empire (247 BCE to 224 AD) and then the Sassanid Empires (226-651 AD).

History documents an awakening of consciousness that occurred throughout the world during what has been termed the Axial Age beginning in the 7th century BCE and it lasted for centuries. During this age, a set of transformational religious and philosophical ideas develop independently around the world. In China Confucianism develops ideals of ethical government and societal conduct, and Taoism develops the philosophical precepts to understanding the way or divine order of existence. In India, Buddhism and Jainism gave rise to the awareness of a higher level of consciousness that can be realized within humans, individually and collectively. This higher awareness brought about relative prosperity and peace in each culture.

In Persia, Zoroastrianism develops a set of high moral ideals and aspirations for the evolution of a higher consciousness and a singularity of God. Judaism also developed a set of moral codes, or commandments, and promoted the idea of a one all-inclusive pervasive deity or God. During this age the Greeks developed an awareness of the Divine through science and philosophy. These philosophical traditions, represented by Socrates, Plato and Aristotle, brought about the ideals of how human beings could live together in greater prosperity, happiness and peace through representation of government and leadership that valued personal freedom. These ideals were diffused throughout Europe and the Middle East in the 4th century BCE by

Alexander III of Macedon, more commonly known as Alexander the Great, whose conquests extended all the way to India in the East and Hellenized the entire Mediterranean, influencing all the cultures of modern Europe that are still infused in Western culture today.

However, as Greek civilization began to dominate and control people of other cultures, it also began to be successively chipped away by rebelling regional powers. It lost its moral compass and fell into decline. China's Han Empire (206 BCE to 220 AD) also fell into civil war as their empire had become so big and included so many ethnicities that it could not be centrally managed. People of different cultures, languages and customs within the empire were disenfranchised and underrepresented. Here we see early on how power in the hands of a few destroys civilization.

By the 3rd century BCE, the Roman Empire developed technologies and an army that enabled them to expand their empire. The ability to make roads that linked key trading areas and mobilize armies, the development of cement and ability to build great structures, and create a set of laws and administration made Rome the marvel of the known world at that time. Even their enemies were enamored, and this drew many to their cause. By the reign of Emperor Augustus (late 1st century BCE), Rome controlled all the lands surrounding the Mediterranean. By the reign of Trajan (early 2nd century CE), Rome controlled much of the land from England to Mesopotamia. Romans, and even those they conquered, developed cohesion from the set of laws and justice that were expounded.

In the 3rd century BCE, most of South Asia was united into the Maurya Empire by Chandragupta Maurya, which flourished under his successor Ashoka the Great. Although Ashoka was a warrior himself, he later became a Buddhist and advocated non-violence, love, tolerance and vegetarianism that enabled his empire to live peacefully and prosper for generations and helped spread enlightenment throughout much of Asia. This ushered in India's golden age, and science, engineering, art, literature, astronomy and philosophy flourished under the patronage of successive kings.

Around 325 AD, the Roman emperor Constantine rallied the poor and downtrodden masses that had adopted the emerging faith of Christianity under his banner, and took control of the region and the religion. This movement for a righteous cause initially enabled Constantine to take the throne and freed many people, but his sole domination and ultimate repression of freedom and rights ended up fracturing his empire which led to internal revolts and invasions by the barbarians. First came the Germanic people from the North, the Goths, Vandals, Lombards, and Franks, and later from the East by the Huns, Avars, Slaves, Bulgars and Alans. The mighty Roman Empire was further weakened by the separation of power between Rome and Constantinople which became known as the Byzantine Empire. This weakening led to further invasions by the Muslim Arabs, then Vikings of the North, then the Muslim Moorish, Turkic and Mongols. This led to what is called the Dark Ages and Middle Ages where science and culture in Europe degraded as a centralized control repressed freedom, liberty and justice.

While Europe was living in the Dark Ages, a new light of hope was dawning in the East with the advent of the Prophet Mohammad, whose divine inspiration for mankind living together in a global brotherhood called Islam spread throughout much of the world. During the Islamic Golden Age (750-1258 CE), Muslim leaders dominated trade and gained power over the Mediterranean and Central Asia. They integrated the skills of the ancient Middle East, of Greece and Persia, and added the innovations of paper from China and the decimal positional numbering from India, and incorporated new innovations such as trigonometry, optics, astronomy and medicine.

Islamic culture became the center of science, philosophy and engineering. But this empire too became too big and centrally controlled to manage, it too became repressive and dominating, and it too began to crumble with successive invasions from the Mongols, the crusaders of Europe, and local uprising which ultimately brought about its decline. The ideals put forward in the *Koran* were usurped by those who used it to maintain power and control over others, just as the Holy Church had done in Europe. So we can also see how religion is often used by a select few to exert control over the masses, and how it corrupts into despotism again and again. Power corrupts, and absolute power corrupts absolutely.

The Middle Ages ended with the advent of the Reformation and decline in the central authority of the Catholic Church, ushering in what is known as the Early Modern period. From the 1500's to the 1800's Western Europe began to rise in importance with the development of new technologies and new republican ideas of freedom, liberty and justice for all. Foremost among the new technologies was the invention of the printing press and with it the dissemination of information and development of free thinking among the masses. This period is characterized by the rise of importance of science and rapid technological progress, secularized civic politics and the development of the nation state (as opposed to kingdoms). This liberal thinking brought about the possibility of mankind living in peace through self-rule and personal liberty where feudalism had been the norm and had kept people in abject poverty and desperation.

Europe's Renaissance in the beginning of the 14th century gave rise to the rediscovery of the classic world's scientific contributions which led to humanism and the Scientific Revolution. During this period European powers came to dominate the world. Europe's maritime expansion, with modern ships and weapons, expanded its geographic control. At first Portugal and Spain were the predominant conquerors who built global empires upon which the "sun never set". But soon the more northern English, French and Dutch elite began to dominate the Atlantic. In a series of wars fought in the 17th and 18th centuries, culminating with the Napoleonic Wars, Britain emerged as the new world power.

During this time the development of debt instruments and international banking enabled a small cartel to control whole nations, their people, politics and natural resources. Here began modern history's epic for world domination, and the fight between royal families, the financial elite, and their bankers to exert control over the masses. Bankers would finance both sides of warring nations and reap sums of money and power never known before without the public ever realizing who was ultimately in control. It was during this period that the first multi-national corporations were formed, such as, the Dutch and British East India companies, and a global cartel began to dominate under the secrecy of the corporate veil. Banks and corporations have taken over the world, and even to this day few realize who is really in control.

The Age of Enlightenment began at the end of the 1700's, and ushered in a period of free thinking and utopian ideals. This movement served as an inspiration to reform society and advance knowledge; it sought to enlighten the consciousness of man. British literary intellectuals such as, John Locke and David Hume, and French intellectuals such as Voltaire and Rousseau, promoted questioning of previously held beliefs, authority and control by the rich and royal families, and advocated free and critical thinking as the basis of creating a new and better world.

The German philosopher Immanuel Kant's (1724-1804) essay "Answering the Question: What is Enlightenment?" brought into question the established conditioning of the masses. According to Kant, enlightenment was "Mankind's final coming of age, the emancipation of the human consciousness from an immature state of ignorance and error". Frederick the Great, king of Prussia from 1740-1786, saw himself as a leader of the Enlightenment and patronized the philosophers and scientists of his court in Berlin. This movement awakened the consciousness of Western civilization to the realization of what it could become, and how mankind could live together in happiness and peace.

The age of enlightenment is perhaps best illustrated in the revolutions of the late 1700's. Free thinkers like Benjamin Franklin, Thomas Jefferson and Thomas Pain were able to envision a more democratic existence and saw republicanism and giving representation and freedom to individuals as the cornerstone of a world being able to live in prosperity, happiness and peace. They saw the totalitarian control through central banks and their governments as the prime cause of war and repression around the world and sought to overcome its grip through first the dissemination and free exchange of ideas, and ultimately by challenging the control over the elite through revolution.

The ideas and ideals of Enlightenment are illustrated in the American Declaration of Independence (1776) and Bill of Rights (1789), the French Declaration of Rights of Man and of the Citizen (1789), and in the Polish-Lithuanian Constitution of 1791. However, even now humanity is still striving to reach its own potential for freedom and peace. Women have had to fight for their rights, minorities have had to fight for their rights, and citizens still have to fight to keep the rights that have been won. The more unconscious a population is the more they are taken advantage of, and conversely the more people question and become conscious of how they are being manipulated the more powerful they become. Freedom, liberty, prosperity and peaceful existence are the direct result of a people becoming conscious – collectively.

The elite still create wars to dominate other nation's resources and suppress the rights of citizens to maintain control over the masses. The international banking system still has control over all world governments through the International Monetary Fund (IMF), World Bank and Federal Reserve cartels, who in turn exert pressure on governments that they control to create laws that favor the elite – working through corporations, holding companies, trusts, and their lobbyists – and then place the burden of taxation on the working man. Both World Wars were created for profit and control of world resources, by faces and names the public rarely sees or hears. Few ever learn the true underlying causes of these wars.

History has taught us, and continues to teach us, one clear lesson: the rich will continue to attempt to dominate the world through the control of the money supply and thereby the control of governments, natural resources, information and the politics that govern a people; and the more power they have, the more the masses suffer. It is just as we have all heard: those that have the gold rule. People need to maintain control of their nation's money supply, natural resources and representative just as Thomas Jefferson, and so many American presidents, have warned, in order to keep their democracy strong and nation prosperous.

The more freedom the citizen has, the greater the flow of information, innovation, science and commerce. The more power the people have, the happier more people are, and conversely the more power the elite maintain the more people suffer. The more conscious we all become, the better off we all are. Raising the consciousness, individually and collectively, brings greater happiness and peace to the world.

Chapter 9 – The Human Brain

Our understanding of our self and our world is related through our mind and senses, so to better understand who we are and our relationship to our world we need to understand how the brain works and what understanding is. From the earliest interactions of energy particles creating systems of organizing intelligence we can observe circuits of awareness forming. Atoms interacting with other atoms form patterns, and these patterns form loops or cycles based on their attractive and repulsive forces as positive and negative charges. As these protons, neutrons and electrons interact little circuits are formed.

A charge runs through the atom and a field of magnetism is created around the atom or atoms. These fields conduct the energy and hold the patterns that create the pulses, vibrations or character of these atomic structures. As more complex systems of atoms develop they form molecular structures, and in humans, amino acids, proteins and cells; it is this current of energy that animates the existence of these substructures within the human. At one end of the cell a predominately positive charge is held, and at the other a predominately negative charge is conducted. Before a cell divides, the chromosomes align according to this polarity.

As cells come together to form more complex life forms the electro-magnetic current of energy is conducted throughout the life form, and all the cells within it are conducting the same field of energy – they share the same field. This current of energy running through the cells of the life form developed into what is called a nervous system. The nervous system is an organ system containing a network of specialized cells called neurons that coordinate the actions of an animal and transmit signals between different parts of its body.

The nervous system in human beings is divided into two parts by science: The central nervous system contains the brain, spinal cord, and retina. The peripheral nervous system consists of sensory neurons, clusters of neurons called ganglia, and nerves connecting them to each other and to the central nervous system.

All these systems are interconnected by means of a complex of neural pathways. The energy that runs through these pathways forms the principal life force that animates our existence. In addition to these neurons which serve as a kind of wiring system for the human body, there are channels of low resistance and high conductivity that channel energy throughout the body that the Chinese call Chi; as well as a field of bio-magnetism conducted around the whole body containing the field imprints of the human body. In these ways, consciousness is conducted throughout the body.

Neurons send signals as electrochemical waves travelling along thin fibers called axons, which cause chemicals called neurotransmitters to be released at junctions called synapses. A cell that receives a synaptic signal can either be excited or inhibited or otherwise modulated. Sensory neurons that are excited from some sense that they have been calibrated to detect, send signals that inform the central nervous system of the state of the body and its external environment. For instance motor neurons being stimulated in the peripheral ganglia in the skin, upon being touched or sensing heat, would send a signal throughout the collective nervous system of the affected body part as well as the central nervous system that would cause the body parts to react. Touching a hot plate might cause the hand to recoil.

The size of the nervous systems in animals ranges from a few hundred neurons in simple life forms like worms, to billions in a human being. The basic design within all animals is to have a paired set of neurons working together forming a circuit. This is called a bilateral nervous system, meaning the left and right sides are approximate mirror images of each other. All bilateral nervous systems are believed to have descended from a common wormlike ancestor that appeared in the Cambrian period, some 550-600 million years ago. The formation of a human fetus begins with such a nervous system.

The fundamental bilateral body is a tube with a hollow cut cavity running from the mouth to the anus with a nerve cord running through it. There is an enlargement or ganglion developed along each body segment providing more nerve cells to develop as motor or sensory nerves, with an especially large ganglion at the front which is called the brain. As an example, earth worms have dual nerve cords running along the length of the body and merging at the mouth and tail. These nerve cords are connected by transverse nerves like the rungs of a ladder. These transverse nerves help coordinate the two sides of the animal. Two ganglia at the head end function similar to a simple brain, while photoreceptors on the animal's eyespots provide sensory information on light and dark.

Senses within animals develop as little circuits. More circuits developed as more senses developed. Motor neurons developed to enable cell structures to move and to feel. Other cells developed the ability to sense molecular interactions as taste and smell, and identify food, predators or mates. Other neurons developed the ability to sense sound waves, and as vibrations were picked up in the cell, it transmitted an electrical impulse to the central nervous system which we call hearing. Cells developed that were sensitive to the energy we call light, and picking up the variants of light and transmitting that signal to the brain we call seeing. As life forms evolved their senses developed as the need arose, or were eliminated if not needed. In humans a sixth sense is developing, which is the sense of self itself.

The evolution of a complex nervous system has made it possible for various animal species to have advanced perception abilities such as vision, complex social interactions, rapid coordination of organ systems, and integrated processing of concurrent signals. In humans, it has made possible the development of language, abstract representation of concepts, emotion and cultural developments. Our sense perception is the interaction of cells within our nervous system, and thoughts and emotions are essentially electrical and chemical impulses running through fields of neurons in our head and body.

Within our brain lie billions of brain cells called neurons. Each neuron looks like a blob with spines sticking out. The little spines or feelers near the center are called dendrites and the longer ones are called axons. Each neuron can be thought of like a little personal computer, and all these little PCs in your brain are connected through an internet system of axons. There is essentially a web of axons and dendrites connecting all your brain cells. Brain cells are like carbon based chips that are connected through a web of tissues that form the mass of your brain. Thought is like energy that runs through the chips in your computer to process a function.

Most neurons send signals via their axons, although there is some dendrite-to-dendrite communication too. Neural signals propagate along an axon in the form of electrochemical waves science calls action potentials, which produce cell-to-cell signals. The places where the axons come together and form a chemical bond to keep them together are called synapses. Synapses send electrical and chemical signals through chain reactions, allowing the charge to run from a series of neurons that are all linked through their synapses. These combinations of connections form the thought patterns in your brain. As thoughts change, the places where the axons come together can reform making new connections and patterns of thought. Your thoughts are simply programs formed in your brain.

The networks formed by interconnected groups of neurons are capable of a wide variety of functions, including feature detection, pattern recognition, and timing. Neurons also create layers to add depth to pattern integration. For instance, in a visual system, the sensory receptors in the retina of the eye are only individually capable of detecting "points of light" in the outside world. A second-level of neurons associated or connected with this basic function can add a hierarchy of processing stages. This enables the sense of vision to include variances in brightness, color, tone, movement and the association of images with behavioral responses.

The human brain is the center of the human nervous system. It is estimated to have upwards of 100 billion neurons contained within it. The most basic functions of sense perception that developed early in animal evolution tend to lie near the center of the brain, while more advanced mental processing grew outwards and upwards expanding the brain in size. Most of the expansion comes from the cerebral cortex, the outer folds of greyish pink matter crumpled up in our skull we think of as our brain. The more developed or advanced processing abilities occur in what is called the frontal lobes, on the very top to forward part of our skull. The frontal lobes are associated with executive functions such as self-control, planning, reasoning and abstract thought and these areas are still developing within human beings.

The adult human brain weighs about 3 pounds and is the consistency of tofu. The size by volume of the brain is around 1130 cubic centimeters in women and1260 cubic centimeters in men. Interestingly, the Neanderthals had a larger brain than do present-day humans. The cerebral cortex is nearly symmetrical, with left and right hemispheres that are approximately mirror images of each other. Scientists divide the brain into four lobes – frontal, parietal, occipital and temporal – corresponding to bones in the skull rather than brain function. Each part of the cerebral cortex stores various skills, memories and functions.

The brain is also divided by functional categories or regions. One consists of the primary sensory areas, which receive signals from the sensory nerves and tracts by way of relay of energy. The primary sensory areas include the visual area of the occipital lobe, located in the back of the head. A second category is the primary motor area, which sends axons down through motor neurons in the brainstem and spinal cord. This area occupies the rear portion of the frontal lobe, basically in the middle of the head. The third primary category of the remaining parts of the cortex is the associate areas. These areas receive inputs from the sensory areas in the lower parts of the brain and involve the complex processes we call perception, thought, and decision. Energy moving through these fields of neurons in your brain creates the sense of you in your own mind.

Different parts of the cerebral cortex are involved in different cognitive and behavioral functions. Motor areas innervating each part of the body arise from a distinct zone in the brain, with neighboring body parts represented by neighboring zones in the brain. Electrical stimulation of one part of the brain will result in a muscle contraction, stimulating another part of the brain will elicit a memory. Associative memories tend to be stored in similar parts of the brain, but the senses related to that memory will be connected to different parts of the brain.

If you think of a duck, one part of your brain will recall memories of being with ducks, and through axons your mind will also connect with the parts of your brain that can visualize a duck, recall the sound of a duck, the feel of a duck, the taste of a duck and perhaps the memory of the cartoon Donald Duck and the emotions associated with that or those images. The memory and association of Mickey Mouse is likely to be found near that of Donald Duck, and the emotions associated with those cartoon characters from your childhood are likely being triggered as you think about them now as you are reading this, causing you to smile.

Each hemisphere of the brain interacts primarily with one half of the body, but they are processed on the opposing side to create a sense of whole. In other words, the left side of the brain interacts with the right side of the body, and the left side of the body interacts with the right side of the brain. The two cerebral hemispheres are connected by a large nerve bundle called the corpus callosum, which draws together the nerve fibers from the left and right hemispheres into a bundle underneath called the thalamus. This relay point is known as the "Cave of Brahma" by the yogis for it draws together all the interactions of the mind that create the sense of Self – Brahman. But there is something other than the mind that observes the mind, and that is the consciousness itself.

The left and right sides of the brain also tend to associate life through different perspectives. The left side is more sequential whereas the right is more simultaneous. The left side is more detail oriented and the right more holistic. The left side is more word predominate and the right more pictorial. The left is more logical and the right more intuitive. The left specializes in numbers the right with shapes. The left develops measurements and the right motions. The left side of the brain more actively recalls the past whereas the right imagines the future. The left focuses on grammar and the right on intonation and emphasis. The left side works with patterns, and the right side with accents. The left is more literal and right more abstract. Each side of the brain works together to form a greater whole in awareness and understanding. The two sides play off each other, self-reflecting, forming a higher state of consciousness than if they worked singularly.

We also find the left side of the brain focuses on the content of what is being assimilated whereas the right side focuses on the context of what is being assimilated. The right side will focus on name recall and the left facial recognition, providing a more complete composite of association. The left side will provide time awareness and the right spatial awareness to create a greater time-space perspective. The left side develops components of something, the right the whole object. The left side develops more the science of understanding and the right the art. Math is more focused in the left and music is more focalized on the right. Each side of the brain work together to create an expanded awareness in our mind.

This development of brain function has led to an expanded awareness in human consciousness. The consciousness that is innate in existence is expressing through ever greater degrees of self-expression and self-reflection. One hundred years ago very few people had developed themselves to the point that they were able or inclined to know who they are or why they were born, but now millions of people are seeking to find themselves and enlighten their consciousness. They are becoming conscious of their own consciousness, and seeking to further evolve themselves to the full enlightenment of the consciousness. We are collectively seeking enlightenment. We are all evolving greater conscious awareness through our life experiences. We are all developing greater self-awareness.

Chapter 10 – The Eastern Perspective of Consciousness

Consciousness in and of itself cannot be rendered into a mental process or human experience. Consciousness is undifferentiated, complete and whole within itself. As soon as we try to define it or even understand it through a mental process or experiment, we create an illusion. All mental processes are self-created, and the thought of consciousness itself is transient, illusionary and ultimately not real. Consciousness itself isn't a thing, and it cannot be reduced.

There is consciousness itself, what we can call God, the underlying essence of all that is, which is to be distinguished from what our mind can be conscious of. Metaphorically, this Divine consciousness can be thought of as the mind of God, although any functions that appear to arise out of it, such as the Big Bang, are in and of themselves only a reflection of the latent potential inherent within it, and not it Itself. However, most of science views consciousness as it is expressed through the mind and subjective experience and as a temporal manifestation or physical expression. So when we discuss consciousness itself it must be within this context and understanding of how the word is being used.

There is an underlying apparent potentiality from which all things arise, are sustained by, and merge back into. The human experience through mind and its senses separates all that is to make distinctions that appear to the mind as reality. We address or experience consciousness subjectively, and to the degree to which our own awareness develops, this enables us to become aware of this reality. To know what lies beyond the mind, one's awareness must transcend the mind.

To gain a deeper understanding as to what consciousness is, and how to realize the true nature of existence, we need to understand how our own mind and senses work. Our mind creates its own sense of self that obscures the reality of our own being. Thus, we need to become aware of what our awareness is and how the mind interprets sensory information and relates it to its own self-created sense of self.

To begin with, we need to reflect on the nature of apparent differentiation, as reflected in what we call the Big Bang and the expression of physical existence. Before anything existed, there was an apparent potential for it to exist. In other words, prior to existence there was no time, space, energy or matter. Nothing was, or ultimately is, separate from anything else in its true intrinsic nature. The potential for anything to arise is created when reflected upon, just as your own thoughts are a reflection of your own state of consciousness. Once reflected, it appears real, if only as a thought, and thoughts lead one to an apparent reality and sense of individual self.

This same self-reflective quality is inherent within all material expressions. Anything that appears separate is only separate in appearance, on the surface, and most of what is, is not appearing. This duality of consciousness is created by a subjective reality that is created within our brain's associative patterns of awareness. Consciousness expresses through everything that is materially expressed, including the mind, but is not materially of the mind or any material expression.

The discussion of consciousness and mankind's search to find or discover it is as old as recorded history itself. Within the literature of the yogis it is referred to as the Sanatana Dharma, the eternal truth. Evidence of this search lies in the images of yogis sitting crossed legged in meditation in the archeological discoveries found at the Mohenjo-daro site, and others, within the Harappan culture and Indus Valley Civilization, now in modern Pakistan, dating back to pre-history. Physical evidence dates back to the Mehrgarh I period over 7000 years BCE. Here Shiva lingums, icons reflecting the procreative energies of consciousness (latent potential) and Shakti (patent expression), are found everywhere. This iconography is indicative of the practice of merging individual consciousness with total consciousness, or the enlightenment of the consciousness to full God realization.

This primordial search to realize the true nature of the Self and God is generally called yoga in the East, typically using some form of meditation or introspection. These truths were studied by the yogis in traditions now referred to as Shaivism (Shiva), Vaishnavism (Vishnu) and Srauta (Vedic) traditions within the Hindu culture. These studies involve transcending mental or cognitive understanding and sense perception. These ideas were eventually transcribed and recorded through a series of works known as the Vedas. Later these spiritual concepts were further expounded in a corpus of literature known as the *Upanishads*, the *Puranas*, the *Sutras* and finally in *Vedanta*.

The earliest known physical writing on the subject is called the *Rig Veda*, and the earliest physical copies date to around 1,700 BCE. Rig is rooted in the Sanskrit word meaning "praise, verse" and veda is "knowledge", and this ancient Indian sacred collection of Sanskrit hymns provides metaphorical and poetical accounts of the origin of the world, hymns praising God in the various aspects to which He appears and various prayers. Here the Divine is worshiped as the primordial fire from which life was created, represented by the deity of Agni, and as the flow of life itself known as Indra. It is from the word Indra that the names Indus River and Indus Valley civilization are derived, and from which the geography known as India comes. Indus was pronounced as Hindus by the Persians who led the Europeans to India, so the Europeans who later colonized India gave the people and culture the name Hindus. The people living in India never thought of themselves as Hindus.

The essence of the Vedas is to first recognize and worship the source of creation itself or God, and to strive to attain knowledge of this Creator and seek knowledge of its, or His, will. The yogi seeks to align the individual will with Divine will. After the *Rig Veda*, the *Yajurveda, Samaveda*, and *Atharvaveda* were composed over successive generations to further expand the knowledge and understanding of the Divine, using deities to represent those attributes and aspects of creation. Within these texts, the concepts of karma (action) and dharma (natural law) were developed with the intention of bringing human will in line with Divine will.

Shivism is rooted in the primordial understanding of non-dualism, or the realization that everything is of the same intrinsic nature, and that everything physically manifest is God being reflected in the multiplicity of appearances. Shiva or Siva, sometimes called Rudra, is given over 1,008 names. Among the earliest known is the proto-Shiva name of An, the same An of the Sumerian culture. Shiva yogis don't put much attention on ideas or concepts, because the very purpose of their practice is to transcend the trammels of the mind and senses in order to realize God.

Most Shivite yogis connect to consciousness through its expression of Shakti – the life force energy that expresses the underlying consciousness of God. As the same energy which causes their individual existence causes all of existence, that connection with the life force enables the individual consciousness to merge with total consciousness. Typically their guru, or teacher, would transmit this consciousness from a higher level than the mind, and by receiving this energy from their teacher in meditation their mental state of consciousness would attune to those states which lie beyond the mind. By concentrating on the life force energy within themselves in meditation, the yogis raise their Kundalini Shakti to the top of the head where it develops their higher consciousness.

The earliest known Shivite literature comes from the *Shiva Rahasya Purana*, which dates to around 5,000 BCE, then the *Svetasvatara Upanishad*, which dates to between 400-200 BCE, followed by the *Siva Sutras*, from the 8th century CE. Within the *Shiva Rahasya Purana*, the saint Ribhu provides a discourse known as the Ribhu Gita. These discourses are among the first to describe the non-duality of existence, now known as the science of Advaita. Here again the emphasis is on the underlying essence and non-transitory state of consciousness. Advaita discriminates between the eternal (nitya) substance and that which is transitory perceived as existence (anitya). The yogis are guided to liberate their awareness from perception and cognition to join in eternal consciousness known as Satchidananda. Only when consciousness is unbounded by mental impulses can the consciousness be liberated and the true nature of the Self be realized. This is why the yogi sits in meditation, to free himself from himself and his own ideas or perception of himself (or herself).

Within the Svetasvatara Upanishad, we learn that the aspiring yogi or student realizes the nature of existence, God and him or herself through the fire of inward turned consciousness – yoga tapas. The emphasis is on quieting the mind and receiving spiritual energy through the teacher or guru who has already attained enlightenment of the consciousness - a sort of trans-thought transference. Through Shivite literature, we learn to turn our awareness around from going outward through sense perception to going inward to the source of awareness itself within us. The focus of concentration breaks the bonds that keep our awareness tied to mental activity and commensurate desires and ideas.

The *Shiva Sutras* of Vasugupta, from the 8ᵗʰ century CE, approach the attainment of enlightenment, not through an understanding of it, but through the Tantric approach of energy. There is an awareness of the presence of the Divine within us animating our own existence that is accessed or realized through this life force energy or Shakti within us. By focusing on this Shakti the aspirant burns away the attachment of mind and senses, which liberates the individualized consciousness to attain full God Consciousness or God Realization.

The other approach used is Jhana Yoga -- to realize the nature of one's own consciousness directly by observing it. By being conscious of one's own consciousness, the consciousness of the individual is awakened. The reflection of "I am Shiva" within the sutras is approached not from the personal sense of I being anything separate from anything else, but realizing that the fundamental and eternal nature of our self is nothing other than what existed before the body appeared in form or remains after the form is destroyed. Shiva's cosmic dance is the creation of everything and destruction of everything from within Himself (the Divine), and all forms created will be destroyed but their fundamental and eternal nature remains the same and unchanging.

Within the schools of Vaishnavism, a similar philosophy of consciousness is understood, but expressed through different means, focusing more on the manifestation of the Divine through physical representations, such as Avatars. Lord Vishu represents the Supreme Deity and it is through His dream that the universe appears. The Rig Veda states: "Just as the sun's rays in the sky are extended to the mundane vision, so in the same way the wise and learned devotees always see the abode of Lord Vishnu". The emphasis of this path to God is on devotion and submission of worldly desires to the will of God and attainment of God Realization. A turning away from worldly desires helps free the mind and develops greater spiritual awareness, not unlike the mystic Jewish, Christian, Muslim or Buddhist traditions.

Foremost in the Vishnava literature is the *Mahabharata* and within it the volume known as the *Bhagavad Gita*, which in its written form dates to around 200 BCE. In this most famous of Hindu literature the transcendent Creator, Vishnu, physically incarnates into the human-like form of Krishna who instructs his disciples in the knowledge of the Self. In this epic, war is used as a metaphor to describe the battle between the illusionary aspects of creation and the eternal nature of the Self.

Through devotion, the aspirant is able to transcend the levels of mind and judgment to realize the natural law (or Dharma) behind all actions (or Karma). In the *Gita*, Krishna explains that "all beings are in me", and instructs his disciple and hero in the story, Arjuna, to abandon all forms of desire and surrender his will to Him. This Karma yoga involves aligning individual will to Divine will, and in developing the devotion to serve only God's will. Such a concept was also later developed within the *Bible* and the *Koran*.

This Bhakti Yoga enables the Divine will to guide the individual will to the perfection of the consciousness. When the presence of God fills the mind of the individual, the aberrations of the mind are displaced just as the light displaces darkness. This enlightenment is God Consciousness or Krishna Consciousness. Thus Bhakti Yoga is keeping ever present the presence of God within the mind and in the course of one's actions,. Common spiritual practices involve the chanting of spiritual names of God to keep the mind focused on God.

A similar philosophy is recounted in the other famous Hindu epic, the sage Valmiki's *Ramayana*, written sometime in the 5th to 4th century BCE. Rama is identified as an incarnation of Vishnu, who comes to bring light to a darkened world living in ignorance. In the *Ramayama*, virtue and justice are depicted as the path to God, and right living as the means of finding true happiness and peace. The essence of this approach is to become more conscious of one's thoughts, words and actions, and acting in accordance with Divine will or Karma Yoga.

By the 8th century CE, yogic philosophy developed into what is called Vedanta (the culmination of the Vedas) which focuses on self-realization as the means of understanding the ultimate nature of God. In the non-dualistic philosophy of Advaita Vedanta, the primordial consciousness or Brahman is the only reality. It cannot be said to have attributes whatsoever, and all appearances are called Maya or illusions. In Vedanta, the yogi focuses on that which lies beyond the illusions created through mind and senses, and comes to realize the non-differentiated state beyond all states.

The arch yogi Adi Shankara (788 – 820 CE) states that "Brahman is the only truth, the spatial-temporal world is an illusion, and there is ultimately no difference between Brahman and individual self." The underlying essence of all that Is, is the same, reflected in the adage "that thou art", meaning there is ultimately no difference between the experiencer and the experienced (the world) as well as the universal spirit (Brahman). This philosophy is also emerging within the field of quantum mechanics. Thus consciousness only appears differentiated when appearing in form, and all forms are transitory and thus ultimately not real.

The other great Yoga philosopher of note was Patanjali who wrote the famous *Yoga Sutras* and lived during the 2nd century BCE. Patanjali's Sutras teach the path of Raja Yoga involving meditation to realize God. The second Sutra of the Yoga Sutras states that "yoga is the absence of mental modification". In other words, realization or union with God occurs through the absence of thought. Only when the mind becomes still like a placid lake can the consciousness reflect upon itself and the true nature of reality be realized. All the yoga postures and breathing exercises that are associated with yoga today were used in preparation for the yogi to sit in yoga tapas –an extended period of meditation and Samadhi (a transcendental state).

Within the Sanatana Dharma of Yogic philosophy, there are then four principal paths or margs of yoga used to liberate the consciousness from the confines

of self-identification to God realization and enlightenment of the consciousness. These four approaches
are: Bhakti yoga - devotion and complete submission
of desire to God; Karma yoga - aligning ones individual
will to divine will through conscious action; Raja yoga
- quieting the mind through meditation and inward
turned consciousness; and Jhana yoga or Vedanta -
introspection or direct awareness. Most yogis integrate all these approaches in their spiritual practice or
sadhana.

Buddhism follows a similar line as that of the yogis, as
Siddhartha Gautama the Buddha himself studied with
the yogis of India to attain his enlightenment. Buddha
was born a prince who left his wealth and power to
become a wandering sadhu or yogi who lived around
500 BCE. The Buddha is famous for going to great
lengths to attain enlightenment, but only attained it
upon the surrender of the doer trying to attain. What
he had come to realize is the true self is already realized and that which was trying to be something was
getting in the way of just being. Buddha's enlightenment was attained after years of sitting in meditation,
and originally his teachings were taught in silence.

Buddha began his teachings in Saranath along the
Ganges River near modern Varanasi, India. From

his oral discourses later transmitted through his disciples known as the Thera's (elders), the earliest Canon of Buddhist precepts developed such as the famous four noble truths: 1) there is suffering, 2) there is a cause of suffering, 3) suffering ends, and 4) there is a way to end suffering. The gist of the teaching involves becoming conscious of one's thoughts, word and actions and abiding in the true nature of the self. The goal is not only the enlightenment of the consciousness from the trammels of the mind and senses (samsaras), but to endeavor to help enlighten the consciousness of all sentient beings. The goal of the Buddhist is Nirvana, which is the extinction or cessation of cravings and ignorance through an awakened or enlightened consciousness.

Different schools of Buddhism practice different methods of enlightenment, but the commonality is attaining Samadhi or liberation through Dhyana or meditation. Through a focalized mind, or mindfulness, the consciousness of the individual transcends mental aberrations and attains liberation (moksha). Buddha was unique in his day for allowing both women and those of lower casts to join in his Satsangs (discourses) and in his meditation. This led to a populous moment that spread throughout Asia. The essence of Buddhism is realizing the transcendental nature of reality and thereby becoming liberated from those mental conditions that invoke suffering in human beings.

The Jain school of Yoga philosophy developed around the 9th to 6th centuries BCE and emphasizes the necessity of self-effort to move the soul towards Divine consciousness and liberation. Any soul that has conquered its own inner enemies and achieved the state of Supreme Being is called a Siddha. The path is one of learning non-attachment to mental constructs or physical appearances. Through the teachings of the Mahavratas or Jain ascetics, one learns to shed the karmic bonds and attain Divine consciousness. Jains believe that the universe is self-regulated by the laws of nature and that life exists in various forms in different parts of the universe including the Earth. They see life in everything and believe one must recognize the Divine presence in all living things, see the unity in the diversity, and align oneself to this presence through thought, word and deed. In other words, one must become completely conscious.

Mystical schools within Judaism, Christianity and Islam have also recognized the transcendent reality of God and sought to attain a realization of God or become conscious of God through various forms of meditation, introspection and prayer. They all have their own forms of yoga, such as the Kabbalah in Judaism, Christian monasticism and Islamic Sufism. All these mystics take time for introspection, contemplation, meditation and prayer to raise their consciousness and realize God's Divine will. They all meditate in one way or another.

There is a universal understanding that God, the Creator and sustainer of life, is present within our own existence and provides for all existence and through meditation and prayer one can become conscious of this Divine consciousness typically expressing itself in what we call spirit or the Holy Spirit. It is man's eternal quest to raise its consciousness to become more aware of who or what God is and what His will is so that humanity may live in happiness and peace. We are all seeking enlightenment in one way or another.

Chapter 11 – The Western Perspective Consciousness

The Western perspective of consciousness tends to be more related to the human experience of consciousness. It is less an underlying aspect of creation itself or God, and more often thought of as a part of the human condition and related to our physical senses and mental activity. Scientists attempt to look within the brain for answers, and see consciousness more in the expression of consciousness through the human condition. Eastern and Western perspectives each have their place and help us gain a greater understanding of ourselves and existence.

Human consciousness is defined by science as the relationship between the mind and the world with which it interacts. It has been defined as: subjectivity, awareness, the ability to experience or to feel, wakefulness, having a sense of selfhood, or the executive control system of the mind. So, by the very methods used in psychology to realize what consciousness is, and by the very presumption of how it is defined, we limit our awareness of its true nature to a subjective experience or physical expression.

In the West the *Bible* provides the underpinnings of our spiritual understanding of God and the nature of existence. The *Old Testament* provides metaphors that reveal an understanding that man intrinsically lives in a communion with his Creator, and that to the degree we listen to His guidance we live in His grace. Our fall from grace came from not listening to God; our discord arises from our ego creating desires or attachments that keep us from following that inner guidance. In *Genesis* Adam and Eve reflect the idea that man lives in paradise when he lives in an awareness of God, and falls from that grace when his mind ignores or denies that Divine guidance by creating selfish desires. The devil is in his head, and his salvation comes from coming to know God and allowing this spiritual presence to be foremost in his awareness and to guide him.

Judaism, Christianity and Islam all recognize the importance of realizing the Creator's will and exercising our intrinsic higher nature through the guidance given through the teachers and testaments provided. The prophets of the *Old Testament*, the teachings of Jesus, and the poetry of Mohammad all address the need for man to submit his own self-created desires to the realization of God's will. It is our duty to become more conscious of this higher level of consciousness to create peace on Earth, and it is our destiny to do so.

Within the Kabbalistic teachings of Judaism, the Age of the Messiah is recognized as a time when the consciousness of humanity is awakened. The goal of mankind is to become enlightened, and only then will peace prevail on Earth. The stories of the *Old Testament* provide insights into the nature of man and how (s)he is to transcend to a higher level of consciousness. Abraham's revelations about there being only one transcendent God, as opposed to many deities, and that human sacrifices (which were common during this age) were no longer necessary to demonstrate one's devotion, as our Creator knows what is in our hearts. Moses' revelations as to the transcendent nature of the Divine and the ways in which man can live in peace, was set forth in the famous *Ten Commandments*. Jesus taught that the Kingdom of Heaven lies within us and worked to create peace on Earth, and Mohammad sought to establish a universal brotherhood of man. Each demonstrates another evolution in the development in human awareness.

There were also other texts used by the priests that were considered too sacred for the lay citizen, but studied by the most learned priests, which also shed light onto the way to develop higher consciousness. In the *Dead Sea Scrolls*, the oldest *Bible* ever discovered, in a codex called the *Book of Secrets* (scroll 4Q299) it is written "How can a man understand without knowledge or hearing? He (God) created insight for His children, by much wisdom He uncovered our ears that we might hear. He created insight for all those who pursue true knowledge (Gnosis) and all wisdom is from eternity". This whole codex explains how God wants us to develop higher knowledge and gain insight or otherwise, to enlighten.

In the *Book of Enoch*, also from the *Dead Sea Scrolls*, we hear stories of Enoch's enlightenment, and how we can to come to the knowledge or realization of God. Throughout the *Zohar*, the principal spiritual text of Kabbalah that was practiced at least as early as the first century BCE before the physical texts appeared during the Middle Ages, the attainment of enlightenment of the consciousness is explicit. According to Rabbi Berg who first translated the *Zohar* into English:

"The Kabbalists teach us that the ultimate purpose of life is to evolve our awareness and consequently a deep hatred towards our errant and self-indulgent ways. When our realization and revelation is far greater than our love for selfish pleasure, then God himself will enter our hearts and our very being will destroy every trace of negativity and selfishness that taints our soul."

Jesus Christ reveals a similar message in the *New Testament* when he states that "the Kingdom of Heaven lies within". It's not some place beyond the clouds or the Big Bang of the physical universe, but rather a state of awareness that resides within us. When he states "I and my Father are one", obviously the physical body of Jesus appearing to the disciples was distinct but in essence or in consciousness they were one. This is a classic statement of enlightenment; Christ was enlightened in the traditional sense and taught his disciples the way to enlightenment.

In addition to the word revealed through the New Testament, many first and second century gospels have been identified that were used by the Semitic Christians who actually walked with Jesus in Galilee, Nazareth and Jerusalem. We know these books were studied by those who actually walked with Jesus and their children and grandchildren in Israel, because fragments of these books have been carbon dated to the first and second centuries and they are referred to through correspondence we have between the first bishops of the Greek and Roman churches of the second and third centuries. These books were rejected by the Greco-Roman church fathers on anti-Semitic and theological grounds (according to them) but attest to their acceptance by the first Christians who actually knew Jesus personally and walked with him in Israel.

These texts, and some are just fragments, include the *Gospel of Thomas*, the *Book of Thomas*, the *Gospel of Mary* (Magdalena), the *Gospel of James* and that of Philip as well as those of the Nazarenes, Ebonites, Egyptians and Hebrews, plus the *Thunder Perfect Mind* and *Dialog of the Savior*. Each of these depicts a teaching that further explains Christ's meaning that the Kingdom of Heaven (or the Father) lay within us. For instance, in the *Gospel of Thomas*, which can clearly be identified as one of the first gospels written and used in the first century, we hear many similar statements which are recounted in the *New Testament* also, but there is one quote absent from the *New Testament* that sheds light on the nature of existence when Jesus' Apostles ask the question "When will the Kingdom come?" Jesus responds:

"It will not come by waiting for it. It will not be a matter of saying 'it is here' or 'it is there'. Rather, the Kingdom of the Father is spread out upon the earth, and men do not see it."

He tells them in another verse that "What you look forward to has already come, but you do not recognize it." In the *Book of Thomas* (the Contender), Jesus tells Thomas "Examine yourself and learn who you are, in what way you exist, and how you come to be.", and later "He who has not known himself has known nothing, but he who has known himself has at the same time already achieved knowledge about the depths of the all". In the *Gospel of James,* Jesus tells James to "know yourself" and tells his brother that the apostles will "be enlightened through me". The message is clear in each Semitic gospel: to look within to find God, to seek self-knowledge and to enlighten the consciousness and attain God realization.

The first Christian monks, called the Desert Fathers, all meditated and sought to develop a higher consciousness and reflected upon the nature of God, existence and themselves. The monastic orders that developed during the middle ages all devoted time for deeper contemplation on the mysteries of the *Bible* and looked within themselves to gain deeper insight to God's will. The reformation was a direct result of Christians questioning the meaning of the *Bible* and they meditated and prayed to develop a greater conscious awareness. By one means or another they were all trying to raise their consciousness and to become enlightened and filled with spirit or Divine grace. They were all seeking enlightenment through Christ's word.

The study of consciousness scientifically in the West dates back to the Greek philosophers who recognized an awareness beyond the human intellect that Plato called Gnosis. Aristotle recognized a higher order and a transcendent reality as well as a consciousness within a human being. Philosophers like Descartes and Locke struggled to comprehend the nature of consciousness and pin down its essential properties.

The French philosopher Rene Descartes (1596-1650 CE) proposed that consciousness resides within an immaterial domain he called *res cogitans* (the realm of thought), in contrast to the domain of material things which he called res extensa (the realm of extension). He suggested that the interaction between these two domains occurs inside the brain, perhaps in a small midline structure called the pineal gland. Interestingly, the yogi's recognize this little endocrine gland as a point of control over the mind, a point they call the "Third Eye".

Other Western concepts of consciousness are broadly divided between two categories: dualist solutions that maintain Descartes's rigid distinction between the realm of consciousness and the realm of matter but give different answers for how the two realms relate to each other; and monist solutions that maintain that there is really only one realm of being, of which consciousness and matter are both aspects.

The English philosopher John Locke (1632-1704 CE) wrote in 1690 the *Essay Concerning Human Understanding* where he defined consciousness as "the perception of what passes in a man's own mind." This essay marks the beginning of the modern Western conception of the self and it greatly influenced Voltaire, Alexander Hamilton, James Madison and Thomas Jefferson. Jefferson quotes Locke in the Declaration of Independence referencing a "long train of abuses".

Locke describes the mind as a blank slate at birth that becomes filled through life experiences and rejected the notion of innate ideas. Locke saw consciousness as evolving from either direct sensory information or through reflection, but he also noted in Book II:

"Thus, from the consideration of ourselves, and what we infallibly find in our own constitutions, our reason leads us to the knowledge of this certain and evident truth, that there is an eternal, most powerful, and most knowing being; which whether any one will please to call God, it matters not."

In 1890, William James popularized the idea that human consciousness flows like a stream. In his book *Principals of Psychology*, he proposes in his "stream of thought" theory that: 1) Every thought tends to be part of a personal consciousness. 2) Within each personal consciousness thought is always changing. 3) Within each personal consciousness thought is sensibly continuous. 4) It always appears to deal with objects independent of itself. 5) It is interested in some parts of these objects to the exclusion of others.

Since the dawn of Newtonian science with its vision of simple mechanical principals governing the entire universe, some philosophers explain consciousness in purely physical terms. Neuroscientists such as Gerald Edelman, Antonio Damasio and philosopher Daniel Dennett seek to explain consciousness in terms of neural events occurring within the brain. Karl Pribram and David Bohm propose a quantum mind theory of consciousness, but these have not yet been proven by any physical experiment. John Eccles argued in his paper "*Evolution of Consciousness*" that special anatomical and physical properties of the mammalian cerebral cortex gave rise to consciousness, and Bernard Baars proposed that once in place "recursive" circuitry may provide the basis for the subsequent development of many of the functions that consciousness facilitates in higher organisms.

Ned Block, an American philosopher of mind makes a distinction between two types of consciousness that he calls phenomenal (P-consciousness) and access (A-consciousness), where phenomenal consciousness consists of subjective experience and feelings and access consciousness consists of that information globally available in the cognitive system for the purposes of reasoning, speech and high-level action control, and that these functions coincide within human beings. Another American philosopher

William Lycan, in his book *Consciousness and Experience* identifies at least eight clearly distinct types of consciousness: organizational consciousness; control consciousness; consciousness of; state/event consciousness; reportability consciousness; introspective consciousness; subjective consciousness; and self-consciousness.

The mystical psychiatrist Richard Maurice Bucke distinguishes between three types of consciousness: Simple consciousness, awareness of the body, possessed by many animals; Self Consciousness, awareness of being aware, possessed only by humans; and Cosmic Consciousness, awareness of the life and order of the universe, possessed only by humans who are enlightened. In Ken Wilber's book *The Spectrum of Consciousness*, he describes consciousness as a spectrum with ordinary awareness at one end, and more profound types of awareness at higher levels.

Dr. David Hawkins' in his books, *Power vs. Force* and the *Eye of the I*, describes consciousness existing at different levels, and he uses a model with a range of human consciousness from 0-1000 and measures or tests these levels using applied kinesiology. Similar tests have been used by measuring the energy levels associated with each state. Hawkins states that each human being, or any idea or concept, has a level from which consciousness is being expressed. Those states of consciousness below 200 are labeled as being negative (e.g. guilt, fear, sadness, anger or pride), and higher states (e.g. courage, love, selfless service, enlightenment, etc.) reside at higher levels.

On this scale of consciousness, most people in Western civilization calibrate somewhere in the 200's, those engaged in spiritual development in the 300's, those with a highly developed mind in the 400's, those with a higher or awakened spiritual awareness in the 500's, while enlightenment occurs above 600. Few ever reach 700's or 800's. According to Dr. Hawkins, our society had made a leap in its collective consciousness during the 1980's and moved from just under 200 to just over 200, and this trend is continuing. He further suggests that by consciously trying to raise your level of consciousness you can do so.

Ultimately, whatever we perceive consciousness to be, that concept is still only a thought and all thoughts are only mechanical processes. Something beyond the mind created and sustains the mind, and to whatever degree we may become conscious through the mind, there is a higher consciousness providing for the mind's very existence. The mind cannot grasp what lies beyond the mind, and to whatever degree we may become conscious that consciousness lies within a potential for ever greater expression.

We live in an age where the consciousness of individuals is growing en mass. Millions of people are becoming more conscious of what their own consciousness is, and striving to become more conscious themselves. The more conscious we become of our own consciousness, the more our consciousness evolves. The more conscious we become, the greater our realization of God, the universe and ourselves. Life teaches us that the more consciousness evolves, the more happiness and peace we experience, and the more clarity and insight we gain, the more purposeful and meaningful our lives become. We are all innately seeking the enlightenment of our own consciousness. The destiny of humanity is to enlighten and live in peace.

Chapter 12 – World Peace

Once we realize that there is something to be realized, our realization has begun. To become more conscious, we need to become more conscious of what consciousness itself is and in what ways we are evolving our consciousness or not. In other words, the more conscious we become of our own consciousness the more our consciousness evolves. The more our consciousness evolves, the happier we are, as individuals and as a society.

Science teaches us that by questioning we learn, and the more we learn about the nature of our existence, the better we live. Through technology and science we have created cures for disease, and created the ability and opportunities to live healthier and more fulfilling lives. History teaches us that by questioning we learn how to live together for the greater benefit of all. The more we learn, the more we grow and evolve and find ways to work together for our mutual benefit.

Questioning brings us closer to answers, and finding answers to our questions is not only intrinsically rewarding but also guides us to a greater common good. Over the span of human existence, we can readily observe that mankind is being innately guided to learn and grow and evolve. We learn from our life experiences, and learning brings about a better person -- a better existence.

Learning and questioning typically occurs in the face of resistance. The great scientific, philosophical, political and social breakthroughs have occurred in the face of opposition, ignorance and even death. But throughout the course of human history, the truth is ultimately revealed and these revelations bring about a better world to live in. Thus, it is necessary, even essential, to question what we have been told, to question authority, and to question ourselves and our own beliefs.

Only in questioning, in thinking critically, and in rethinking our positions can we gain greater perspective and an expanded awareness. The more we use our brain the better it works, and the more we evolve our conscious awareness the more conscious we become. The more conscious we become the happier and more fulfilled we are, individually and collectively.

Man's eternal quest is to come to a realization of who or what our God is, how our universe works, who we innately are, and what role we play in existence. To the degree we discover the true nature of our self, we find the purpose and meaning to our existence. In order to fulfill our purpose and destiny in life, we must first come to realize who we innately are. We must question, we must inquire and reflect deeply, we must develop ourselves in order to further evolve. We ultimately don't find true meaning and purpose in our lives until we realize the true nature of our Self.

It is mankind's destiny to enlighten. We have been evolving to this end from the dawn of time. In the last few generations human being have come to a critical point in the evolution of human consciousness where we are, en masse, seeking enlightenment. The invisible hand of God is guiding us all to a greater realization of God and of our existence. Virtually all the discoveries described in this book are the result of that inner guiding nature directing us to further evolve ourselves.

The next question we need to ask is "to what end?" What is the purpose of it all? What is the purpose of our existence? Within each human being lies the answer. It is imbued within the innate intelligence that is guiding us to evolve. Ultimately, we all want to live in a peaceful and happy world. We innately know right from wrong and intrinsically want to do what brings about a greater good, despite what mental programming might sway us otherwise.

We have come to the point in the evolution of human consciousness where we are ready to realize who we are and why we were born, so that we can fulfill the purpose of our existence and fulfill the destiny of mankind to live in peace and prosperity on Earth. All it takes is the collective intention of the people and their leaders, and the mobilization of human resources to do so.

There is an awakening occurring within the collective consciousness of humanity. By the very fact that you are reading this book, it suggests that you have evolved yourself to the realization of this vision and the part you play in manifesting it. Millions of people around the world are working to awaken and evolve greater conscious awareness. At no time in the history of the world have so many people shared this vision or applied themselves to this goal. A critical mass is forming, and you are the proof.

The greatest shifts in human civilization did not occur due to the acts of a few, but due to the acts of many coming to a similar conclusion and sharing a common goal or vision. Our political, religious and industrial leaders invoke the changes that need to be made only through the demands of the populations that they serve. It takes individuals coming to a common vision and understanding to invoke the changes that need to be made. It takes individuals wanting to make those changes, and then doing so.

An opportunity is at hand. With the awareness that is growing within you now, we have only to further apply what has been learned and share it with others. One match lights the fire, and one fire can illuminate the room. The more we enlighten ourselves, the more light we shine so others may see. The more we all see, the more the world enlightens. We have entered an Age of Enlightenment and you have been called to help bring it about. Your role is to share this awareness with others; to help bring this light into the world. May the whole world live in happiness and peace.

About the Author

Steven S. Sadleir began his quest to find God, and realize the nature of existence and himself as a child. At sixteen he climbed a mountain to meditate and pray that God reveal these answers to him, and began writing his first book *Looking for God* which is a compendium of all the spiritual and religious teachings of the world that became a best-seller after twenty years of research and writing.

After receiving a B.Sc. in Business Administration from Menlo College he continued on to the University of Wales in the United Kingdom on a Rotary Scholarship to study Financial Economics, and specifically International Banking in order to understand who controls the world. He initially worked as an economist for the United States government at the East-West Center and later for Lloyds bank in the Los Angeles office, and later served as a consultant to the banking industry, and investment banker and fund advisor. He then retired early to write and teach.

Concurrent with his financial career he also travelled the world studying with dozens of spiritual masters from most of the major spiritual traditions. Steven is recognized as a Kundalini Master in the lineage of Vethathiri Maharishi after apprenticing with him for many years, and as a Siddha yogi in the lineage of Shivabalayogi Maharaj after completing yoga tapas in 1990. He has since initiated thousands into Kundalini and Shaktipat meditation.

Steven founded the Self Awareness Institute in 1985 and it has since grown to thousands in over 120 countries. He has written numerous books on spiritual, philosophical and economic subjects, and hosts *Enlightenment Radio* on the Internet. He is a professional speaker and frequent guest on radio and television programs. Steven conducts seminars and retreats around the world as well as distance learning programs from his home in Laguna Beach, California.

For more information contact www.SelfAwareness.com

Looking for God, a Seeker's Guide to Religious and Spiritual Groups of the World.

From AA to Zoroastrianism, Christ to Krishna, *Looking for God, A Seeker's Guide to Religious and Spiritual Groups of the World* provides an overview of hundreds of the world's religious denominations, including: Arcane and tribal teachings, metaphysical and New Age groups and dozens of enlightened spiritual masters, televangelists and new thought leaders.

From the back woods of Borneo to Himalayan heights, Steven Sadleir has traveled the world to meet each group and study each path to God to provide you an up close and personal account of the most interesting and controversial religious and spiritual groups of the planet. Now in its 3rd edition with photographs. This book was an Amazon best-seller. On sale now on Amazon.com.

Self-Realization by Steven S. Sadleir

We live in an age where millions of people are being guided to look within and realize who they are and why they were born. Self-Realization serves as a guide book to look deep within and experience the spirit and recognize the soul in each of us.

We each innately know that the answers lie within us, that happiness and peace are to be found within, and that we each have a destiny to fulfill, but how to find this inner knowing? Self-Realization provides the answer. This book is a meditation and a process of awakening occurs within the reader as they are reading it. By America's foremost Kundalini master, learn more at www.SelfAwareness.com This book is available at Amazon.com.

Christ Enlightened, the Lost Teachings of Jesus Unveiled by Steven S. Sadleir

Over the past few decades more information has been dug up out of the desert about early Christianity than exists in the entire New Testament. Steven S. Sadleir spent over 30 years compiling this research to provide the reader a concise summary of these new findings, including summaries of: The Apocrypha, Merkabah, Kabbalah, the Aramaic Bible (Peshitta), the Dead Sea Scrolls, the Nag Hammadi Library and writings of the Church Fathers.

Steven summarizes what's in the first and second century books used by the Semitic Christians in Israel, including: The Gospel of Thomas, the Book of Thomas, the Gospel of Mary, The Gospel of Philip, The Gospel of James, the Gospels of the Nazarenes, Ebonite's, Hebrews and Egyptians as well as the Dialog of the Savior and The Thunder Perfect Mind. These fascinating discoveries shed new light on the teachings of Jesus Christ. This book is available at Amazon.com.

Money & Power, the Secret History

Steven S. Sadleir worked as an economist for the U.S. government, and international banker for Lloyds bank, and investment banker and fund advisor during the .com book and revels what is really going on with the economy and who really controls the world.

Money & Power explains: The creation and control of money, the founding fathers original vision of banking, how the international banking system works, who created the credit crisis and why, what the new world order and globalization mean, who's behind the Federal Reserve, conspiracy theories, the oil industry, and how to turn our economy around. This book is available at Amazon.com.

The Awakening, an Evolutionary Leap in Human Consciousness by Steven S. Sadleir

There is an awakening occurring within the collective consciousness of humanity. Human beings are being innately guided to a higher level of awareness, we are instinctively being guided to evolve a sixth sense and realize our true nature and purpose. *The Awakening, an Evolutionary Leap in Human Consciousness* provides the insights to explain this awakening and how to access this inner knowing.

This book was written after a period of Yoga Tapas, where Steven was sitting in cross legged meditation for 23 hours a day for forty consecutive days and nights in while in India. Steven is the founder of the Self Awareness Institute in Laguna Beach, California. See www.SelfAwareness.com. This book is available on Amazon.com

www.ingramcontent.com/pod-product-compliance
Lightning Source LLC
Chambersburg PA
CBHW051458170526
45166CB00001B/298